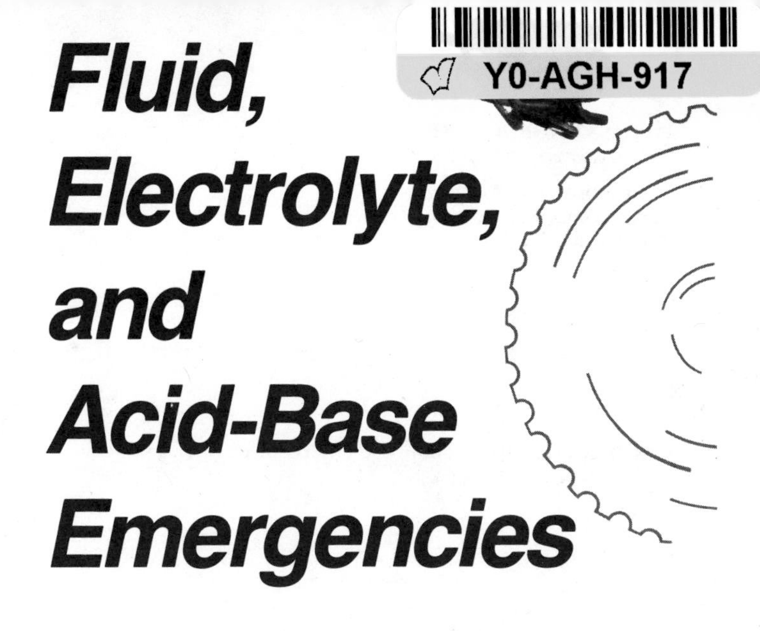

Fluid, Electrolyte, and Acid-Base Emergencies

M.L. HALPERIN, MD

Renal Division,
St Michael's Hospital
and
Professor of Medicine,
University of Toronto

MARC B. GOLDSTEIN, MD

Renal Division,
St Michael's Hospital
and
Professor of Medicine,
University of Toronto

1988

W.B. SAUNDERS COMPANY
Harcourt Brace Jovanovich, Inc.

Philadelphia London Toronto Montreal Sydney Tokyo

W. B. SAUNDERS COMPANY
Harcourt Brace Jovanovich, Inc.

The Curtis Center
Independence Square West
Philadelphia, PA 19106

Library of Congress Cataloging-in-Publication Data

Halperin, M. L. (Mitchell L.)

Fluid, electrolyte, acid-base emergencies/M.L. Halperin, Marc B. Goldstein.

p. cm.

1. Water-electrolyte imbalances. 2. Acid-base imbalances. 3. Medical emergencies. I. Goldstein, Marc B. II. Title. [DNLM: 1. Acid-base imbalances. 2. Emergencies. 3. Water-electrolyte imbalance. WD 220 H198f]

RC630.H34 1988 616.3′9—dc19 87–26650 CIP

ISBN 0–7216–1720–4

Editor: John Dyson
Designer: Maureen Sweeney
Production Manager: Bill Preston
Manuscript Editor: W. B. Saunders Staff
Illustration Coordinator: Lisa Lambert
Indexer: Susan Thomas

Fluid, Electrolyte, Acid-Base Emergencies ISBN 0–7216–1720–4

Last digit is the print number: 9 8 7 6 5 4 3 2 1

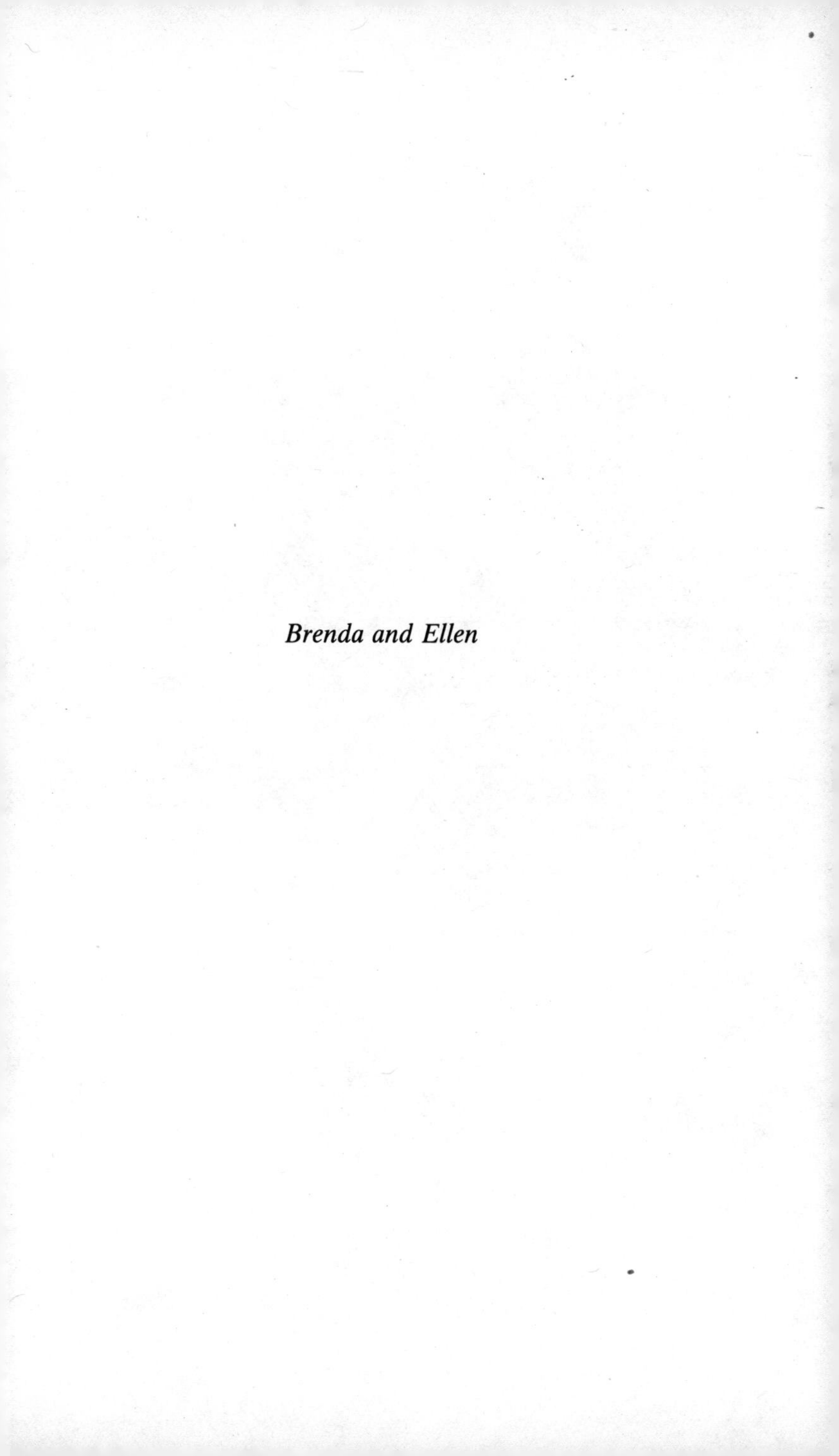

Brenda and Ellen

PREFACE

This book was written with several purposes in mind:

1. To provide a simple and practical approach so that the physician can confidently and safely manage patients presenting with fluid, electrolyte, and acid-base emergencies.
2. To provide the essential physiological and biochemical information so that the reader can utilize the principle of "learn by understanding" and integrate physiology and biochemistry with clinical medicine.
3. To provide information that simultaneously serves the needs of readers with very different levels of expertise. The material is presented to be read at several different levels, based on the needs at the time, without interrupting the flow of ideas.
4. To provide a readily available distillation of critical information to help the physician manage an acute medical emergency.

To achieve these purposes, we have divided the text into two separate but related components. First, there is a continuous text that is located only on the left-hand pages. Second, we have used the right-hand pages in a discontinuous manner to summarize essential material that appears in basic physiology texts, to answer questions asked in the continuous text, to expand on material that is interesting and/or new, to illustrate specific points with simplified diagrams, and, finally, to provide quantitative data to aid comprehension. The additional facts on the right-hand pages are called "Sidelights" and appear in slightly different typeface.

We have three different methods to highlight material in the continuous text. First, there is an overview of the subject presented in the shaded boxes entitled "The Big Picture." Second, there are important, succinct synopses preceding most sections; these are in boxes with a bold outline. Third, there are questions or illustrative cases with questions that are enclosed in boxes with a thinner outline.

Our general approach is as follows: Each major theme begins with an introductory physiology and biochemistry chapter containing the background information necessary to understand the basis of and treatment for each clinical disorder. The subsequent chapters on this theme are more clinically oriented and are designed to enable the reader to deal competently and confidently with patients who have abnormalities in their acid-base balance or fluid and electrolyte metabolism. Clinical examples and flow charts are used liberally throughout the book, and the readers are given many opportunities to test their knowledge.

We offer the following suggestions to aid the reader:

1. Gain an overview of the theme by reading "The Big Picture."
2. Use the illustrative case to establish your level of expertise. The reader who can answer the questions should proceed rapidly through the chapter by reviewing the synopses in the boxes with a bold outline and answering the questions in the boxes with a thin outline.
3. Each chapter outline can be read as a "road map" to the approach to a given disorder discussed in that chapter.
4. Pertinent review articles are listed in each chapter.
5. During an emergency, go directly to Chapter 14. This "Emergency Treatment" chapter provides a basic summary of the issues to be considered in the approach for each of the individual disorders. The physician faced with an electrolyte or acid-base emergency should go to this chapter for guidance in the investigation and management of that case.

We also provide in the book "Clinical Pearls" that contain critical information based on our personal experiences; we discuss highly important issues with respect to history, physical examination, laboratory analysis, and treatment considerations; specific danger signals are emphasized as well.

Chapters 12 and 13 contain challenging cases and additional flow charts that exemplify the integration of the necessary foundation of biochemistry and physiology upon which the clinical decision-making rests (as illustrated on the cover).

ACKNOWLEDGEMENT

We are very grateful to our many colleagues who contributed to the preparation of this book. We especially appreciate the efforts of Linda Peterson, Karl Skoreck, Harold Sonnenberg, Michael West, and Sandy Collins and his staff. We are especially indebted to our good friend David Levine, who was instrumental in initiating this endeavour.

CONTENTS

1

ACID-BASE PHYSIOLOGY

THE BIG PICTURE

The $[H^+]$ is maintained at a very low level and with little fluctuation (40 nM in the ECF and 100 nM in the ICF) despite a very large daily flux (net intake 70,000,000 nmol H^+ per day).

H^+ Input: **Metabolism of the North American diet produces a H^+ load of 1 mmol/Kg/day (exceeds the ECF H^+ content by 10^6-fold).**

H^+ Buffering: **This H^+ load is buffered: 40% by ECF HCO_3^- and 60% by ICF buffers. Buffering is a temporary stage; the H^+ must ultimately be removed. The lungs are an integral part of the ECF buffering process, removing CO_2 and shifting equation (1) to the RIGHT.**

$$H^+ + HCO_3^- \rightleftharpoons H_2CO_3 \rightleftharpoons H_2O + CO_2 \quad (1)$$

New HCO_3^- Formation: **The kidney regenerates the consumed HCO_3^- largely by the metabolism of glutamine and the excretion of NH_4^+.**

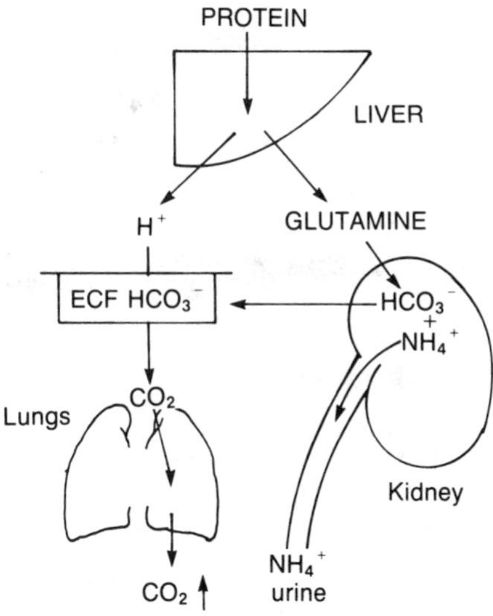

Figure 1–1. Integration of liver, lungs, and kidneys in acid-base balance.

OVERVIEW OF ACID-BASE HOMEOSTASIS

Acid-base homeostasis involves the integrated action of 3 organs: LIVER, LUNGS, and KIDNEYS. The LIVER metabolizes protein, producing 1 mmol H^+/kg daily (70,000,000 nmol/day). The LUNGS remove CO_2 and thereby H^+, allowing HCO_3^- to function as an excellent ECF buffer. The KIDNEYS generate 70,000,000 nmol of HCO_3^- per day, restoring the HCO_3 buffer system to its original state.

ROLE OF THE GASTROINTESTINAL TRACT IN ACID-BASE BALANCE

The stomach secretes about 150 mmol of H^+, per day; this adds an equivalent load of HCO_3 to the body (Fig. 4–1). The small intestine and the pancreas secrete more HCO_3 than this to alkalinize luminal fluid. Normally, that HCO_3 which does not neutralize HCl is reabsorbed; however, disorders in which these secretions are pooled or leave the body induce acid-base disorders.

Intestinal bacteria can modify nutrients and induce acid or base loads. For example, the conversion of neutral cellulose to organic acids leads to an acid load, whereas the complete oxidation of dietary organic anions (which the host cannot metabolize) yields an alkali load. Finally, during malabsorption, the potential acid or base load from the diet can change. Hence, the contribution of the GI tract to acid-base balance can also be very important, especially in disease states.

Essentially H^+ loss is equivalent to HCO_3^- regeneration.
Clinical Application: **The acid-base status of an individual is described in terms of $[HCO_3^-]$, PCO_2, and $[H^+]$. These 3 parameters are linked via the Henderson equation.**

$$[H^+] = 23.9 \times \frac{P_{CO_2}}{[HCO_3^-]} = \frac{\text{Lungs}}{\text{Kidneys}} \quad (2)$$

ILLUSTRATIVE CASE

A 70-kg insulin-dependent diabetic did not take insulin for 2 days. The relevant blood tests were

$[H^+]$	60 nM	P_{CO_2}	25 mm Hg
pH	7.22	$[HCO_3^-]$	10 mM

Questions (answers on page 5)

1. Are you sure that H^+ have accumulated?
2. How many H^+ were buffered in the ECF and the ICF? (Assume no change in ECF and ICF volumes.)
3. By which routes can the body dispose of the H^+ load (regenerate the consumed HCO_3^-)?
4. Has the respiratory system responded to the acidemia? What would the $[H^+]$ be if there had been no respiratory response?

PARAMETERS OF ACID-BASE ASSESSMENT

$$[H^+] = 23.9 \times P_{CO_2}/[HCO_3^-]$$

- $[H^+]$ IS EASIER TO MANIPULATE AND COMPREHEND THAN pH
- $[HCO_3^-]$ REPRESENTS THE "METABOLIC" OR THE RENAL COMPONENT
- P_{CO_2} REPRESENTS THE "RESPIRATORY" COMPONENT
- HCO_3^- LOSS IS EQUIVALENT TO H^+ GAIN

$[H^+]$ and pH

The $[HCO_3^-]$ and the P_{CO_2} together determine the ECF,

Answers

1. Yes: Because the ECF $[H^+]$ increased to 60 nM (normal = 40 nM), and the $[HCO_3^-]$ decreased to 10 mM (normal = 25 mM).
2. In a 70 kg person, the ECF volume is 14 L. The $[HCO_3^-]$ has fallen 15 mM (from 25 mM to 10 mM). Therefore, the total H^+ load buffered in the ECF is 15 mmol/L × 14 L = 210 mmol. Approximately 40% of an acid load is buffered in the ECF; therefore, 315 mmol were buffered in the ICF ((210/.40) × .60 = 315 mmol) and the total H^+ load was 525 mmol.
3. The routes by which a H^+ load are disposed of are
   A. ICF buffering, a temporary measure.
   B. ECF "buffering," resulting in HCO_3^- consumption, which is augmented by hyperventilation (removing the CO_2 displaces Equation 1 (page 2 to the right).
   C. Metabolism of the β-hydroxybutyrate anion to CO_2 consumes some of the H^+ load, regenerating HCO_3^- (Equation 4, page 8).
   D. Renal new HCO_3^- generation is a slower process, but in 3–4 days it will have the capacity to generate 300–400 mmol of HCO_3 per day.
4. The P_{CO_2} is reduced from 40 to 25 mm Hg, indicating hyperventilation is present in response to the acidemia. If hyperventilation were absent, the P_{CO_2} would be 40 mm Hg. The $[H^+]$ = 23.9 × 40/10 = 96 nM (as compared with 60 nM; pH of 7.02, as compared with 7.22).

pH vs $[H^+]$

Because the $[H^+]$ is a very small quantity, it was considered easier to express it as its negative log, the pH. However, in addition to being cumbersome for those not mathematically inclined, the negative log transformation may obscure the magnitude of a change in $[H^+]$; e.g., it is evident when the $[H^+]$ goes from 40 to 80 nM that there has been a doubling of the $[H^+]$, but this may not be as evident when a pH changes from 7.40 to 7.10.

There are several ways in which the physician presented with a pH can determine the corresponding $[H^+]$, and these are discussed in Chapter 2, page 29.

$[H^+]$ and their relationships are shown in the Henderson equation (Equation 2). Note that the $[H^+]$ is depicted in terms of a respiratory component (P_{CO_2}) and a renal component (HCO_3^-). The normal values are $[H^+]$ 40 nM, P_{CO_2} 40 mm Hg, and $[HCO_3^-]$ 24 mM. The $[H^+]$ is often reported as its negative log, the pH. The pH transformation does not improve the understanding of acid-base physiology and, for some, makes it more difficult. Therefore, we shall talk in terms of $[H^+]$, and we encourage the reader to become familiar with this terminology, but we shall also provide the corresponding pH values. Conversion from pH to $[H^+]$ is discussed in Chapter 2, page 29.

Bicarbonate (HCO_3^-)

The ECF $[HCO_3^-]$ is an important determinant of the acid-base status and is referred to as the "metabolic" component of the bicarbonate buffer system. The $[HCO_3^-]$ can be determined in two ways—either calculated from the $[H^+]$ and P_{CO_2} (determined by electrode), using the Henderson equation (Equation 2, page 4) or inferred from the total CO_2 (TCO_2) determination, which is part of the "routine electrolytes" (page 7).

When HCO_3^- loss occurs for whatever reason, the $[H^+]$ increases, owing to displacement of the HCO_3^- buffer equilibrium to the left (Equation 1, page 2). In addition, the P_{CO_2} can be viewed as a near constant, owing to its very large turnover rate.

P_{CO_2}

The P_{CO_2} is the partial pressure of CO_2 in the arterial blood and is the respiratory component of the bicarbonate buffer system. CO_2 is continuously produced by metabolism (10 mmol/min), and an equal quantity is removed by the lungs. (Compare the total amount of CO_2 (1.2 mmol/L) to the flux through the CO_2 pool, which is 2-fold higher per minute.) Thus, in hypoventilation, the P_{CO_2} rises rapidly, whereas it falls rapidly during hyperventilation.

There are two processes that lead to an ACIDEMIA (reduced $[HCO_3^-]$ or increased P_{CO_2}) and two that lead to an ALKALEMIA (increased $[HCO_3^-]$ or reduced P_{CO_2}). Thus, there are 4 primary disturbances of acid-base balance. Metabolic disturbances may coexist with respiratory disorders; this is covered in detail in later chapters.

Total CO_2 vs $[HCO_3^-]$

The $[HCO_3^-]$ can be determined from the $[H^+]$ and the P_{CO_2}, using Equation 2 (page 4). In addition, as part of the "serum electrolytes," many laboratories report the TOTAL CO_2 CONTENT (TCO_2), which is an indirect measure of the $[HCO_3^-]$. The TCO_2 represents the $[HCO_3^-]$ (added acid releases all the CO_2 from HCO_3^-) plus the dissolved CO_2 and carbamino CO_2. In all cases, the dissolved CO_2 and the carbamino CO_2 are so small relative to the $[HCO_3^-]$ that the TCO_2 can be used interchangeably with the $[HCO_3^-]$.

If one has measurements of both electrolytes and blood gases, one should compare the TCO_2 with the $[HCO_3^-]$ calculated with the Henderson equation and ensure that they are similar. If they are very discrepant ($>10\%$), at least 1 of the 3 measured values is incorrect, and the tests should be repeated if the error is sufficient to change the diagnosis or the therapy.

A systematic error (e.g., TCO_2 lower than $[HCO_3^-]$) may result in patients with severe hyperlipidemia. This occurs because the TCO_2 measures the total CO_2 content in a given volume of serum, but the HCO_3 is distributed only in the aqueous phase of the serum. Hence, just as in pseudohyponatremia, a falsely lower value is reported by the laboratory during hyperlipidemia (see pseudohyponatremia, Chapter 7).

A second way that one might fail to solve Equation 2 accurately would be if the pK^1 varied (and some have claimed that this occurs in sick patients). The pK^1 is included in the constant of 23.9 in Equation 2. However, the pK^1 is a physical constant that should change only with changes in temperature or ionic strength.

Miscellaneous Parameters

There are several other derived parameters that have been introduced that also add nothing to our understanding of acid-base physiology, and we therefore shall not refer to them (e.g., standard bicarbonate, base excess, and whole-body buffer base. The interested reader is referred to the critique by Schwartz and Relman on this topic).

GENERAL FEATURES OF ACIDS, BASES, AND BUFFERS

- ACIDS ARE H^+ DONORS, AND BASES ARE H^+ ACCEPTORS
- BUFFERS MINIMIZE THE CHANGE IN $[H^+]$ UPON H^+ OR OH^- ADDITION
- THE $[H^+]$ REMAINS REMARKABLY STABLE AT VERY LOW CONCENTRATIONS (40 ± 2 nM)

Acids and Bases

An ACID is a compound that is capable of PROTON (H^+) DONATION, whereas a BASE is any compound that is capable of PROTON ACCEPTANCE.

An acid (HA) dissociates, yielding the H^+ and its conjugate base (anion, A^-, Equation 3 below). While chemists divide acids into strong and weak ones on the basis of their dissociation constant, this differentiation is of little importance biologically, because at a pH of 7, the dissociation of both types of acid is greater than 99%.

$$HA \rightleftharpoons H^+ + A^- \qquad (3)$$

Clinically, rather than the strength of the acid, the major consideration is the magnitude of the acid load. Several other aspects of the acids are also of biological importance; for example, the ability of the anion to act as a potential "partner" to remove H^+. Examples are discussed below.

1. Can the anion (conjugate base) be metabolized to neutral end products and, in so doing, consume a H^+? For example, betahydroxybutyrate metabolism to CO_2 and water:

$$C_4 H_7O_3^- + H^+ + 4.5O_2 \rightarrow 4\ CO_2 + 4\ H_2O \qquad (4)$$

2. Is the pK of the acid high enough so that, at the pH of the urine, it can be excreted along with a bound proton (e.g., $H_2PO_4^-$, rather than $HPO_4^=$ and H^+)?

Buffers

A buffer minimizes the change in $[H^+]$ when either a strong acid or a strong base is added. It is most effective in defending the $[H^+]$ against acids or bases when it is 50%

DIETARY PRODUCTION OF H^+

Dietary Sources of H^+

1. Sulphur-containing amino acids
2. Cationic amino acids (lysine$^+$, arginine$^+$)

Dietary Sources of OH^- (represent H^+ removal)

1. Anionic amino acids (glutamate, aspartate)
2. Organic anions (e.g., acetate or lactate)

Under normal circumstances, carbohydrate and fat, which are neutral, are metabolized to neutral end products and thus have no acid-base impact. However, during abnormal conditions, these precursors may also be metabolized to organic acids in what is referred to as "metabolic acid production":

HYPOXIA: glucose → 2 H^+ + 2 lactate$^-$ (5)
INSULIN DEFICIENCY: triglyceride → H^+ + β-hydroxybutyrate$^-$

Quantitative Aspects of H^+ Overproduction

Hypoxia. If no O_2 were delivered to tissues, 3600 mmol of lactic acid per hour would be produced if there were sufficient glucose to meet ATP needs. Since the total buffer capacity of the ECF is 375 mmol (25 mM HCO_3 × 15 L), and this represents about 40% of the total (close to 940 mmol), death would occur after minutes of anoxia. To titrate this acid load with $NaHCO_3$, 60 mmol of $NaHCO_3$/min must be given (in 25 min, you would double the ECF Na content). Since the kidney at best can only make 0.3 mmol of HCO_3/min, survival depends on the absence of total body anoxia, and the best therapeutic leverage is to stop H^+ production by restoring O_2 delivery. $NaHCO_3$ therapy only buys very little time.

Insulin Deficiency. In the absence of insulin, the maximum rate of β-hydroxybutyric acid production is 100 mmol/hr. Thus, the clinician has 4–5 hours before half of the 940 mmol of normal buffer capacity is exceeded. Long-term survival depends on stopping H^+ production with insulin, but acute HCO_3 administration may be of value before insulin can act in very severe diabetic ketoacidosis. The kidney is not critically important as a source of HCO_3 because it can only generate 10–20 mmol of HCO_3/hr.

Protein to H_2SO_4 and HCl. This is a relatively low rate of H^+ production (i.e., about 4 mmol/hr). Thus, many days are required to diminish the buffer capacity, even during renal failure. Normally, the kidney prevents metabolic acidosis, as its rate of HCO_3 formation exceeds this H^+ load by several fold. However, an elevated plasma [H^+] is required to achieve the maximum renal response. Finally, this type of metabolic acidosis is easily handled by exogenous $NaHCO_3$. The dose required is as much as 70 mmol/day to match dietary H^+ input.

dissociated, i.e., having an equal amount of undissociated acid and its salt. The pH at which an acid or base is 50% dissociated is known as its pK. The effectiveness of a given buffer is determined by:

1. Its concentration (the quantity of H^+ that can be removed).

2. The $[H^+]$ required to achieve H^+ binding: i.e., its pK relative to the $[H^+]$ in the compartment in which it is to be active.

The major buffers in each compartment are as follows (Table 1–2): HCO_3 in the ECF, histidine in proteins and dipeptides, and HCO_3 in the ICF, whereas phosphate is the main urine buffer (NH_3 is a H^+ trap rather than a true buffer) (page 23).

3. The possibility for acid removal in the form of a volatile acid derivative (e.g., CO_2, page 13).

Question (Answer, page 11)

5. Red blood cells produce 200 mmol of lactic acid per day. Why doesn't this cause severe acidemia?

HCO_3 BUFFER SYSTEM: THE ECF BUFFER

The HCO_3^- buffer system is the major ECF buffer, and the assessment of its components provides the basis for the clinical evaluation of the patient's acid-base status. The ECF $[H^+]$ is a function of the ECF $[HCO_3^-]$ and the P_{CO_2} as is evident from the bicarbonate buffer system equilibrium.

$$H^+ + HCO_3^- \rightleftharpoons H_2CO_3 \overset{CA}{\rightleftharpoons} H_2O + CO_2 \qquad (1)$$

Acidemia (an increase in $[H^+]$) occurs when there is either a fall in $[HCO_3^-]$ or an increase in P_{CO_2} (both displace the equilibrium of Equation 1 to the LEFT, generating additional H^+). Conversely, alkalemia (a fall in $[H^+]$) occurs when there is either an increase in $[HCO_3^-]$ or a fall in P_{CO_2} (both of which displace the equilibrium to the RIGHT).

The $[H^+]$, $[HCO_3^-]$ and P_{CO_2} are thus the 3 parameters reflecting the acid-base status of the ECF (the HCO_3^- buffer system is in equilibrium with the other body buffers and thus reflects their degree of titration).

TABLE 1–1. Definition of Acid-Base Disorders

Condition	$[H^+]$ (nM)	$[HCO_3]$ (mM)	P_{CO_2} (mm Hg)
Normal*	40 ± 2	25 ± 2	40 ± 2
***Acidosis*†**			
Metabolic	Higher	**Always lower**	Lower
Respiratory	Higher	Higher if chronic	**Always higher**
***Alkalosis*†**			
Metabolic	Lower	**Always higher**	Somewhat higher
Respiratory	Usually lower	Lower if chronic	**Always lower**

*Normal values are provided for reference purposes. The primary change is indicated by the bold print.

†The term *acidosis* means a process that tends to increase the $[H^+]$. If the blood $[H^+]$ is actually elevated, we use the term *acidemia*. The converse is true for *alkalosis* and *alkalemia*.

Answer

5. 200 mmol is a substantial acid load, and it produces an acute, temporary metabolic acidosis. However, it does not cause a *sustained* acidemia because the lactate anion is delivered to the liver, where it is metabolized either to glucose or to CO_2. Metabolism of 200 mmol of the lactate anion to neutral end-products consumes the 200 mmol of H^+.

In contrast, in LACTIC ACIDOSIS associated with tissue hypoxia, the rate of lactic acid production is considerably greater than the rate of liver lactate metabolism; hence, the $[H^+]$ may rise to dangerously high levels (see page 9).

THE ICF BUFFERS

Sixty percent of an acid load is buffered in the ICF. The major ICF buffer is the imidazole ring in HISTIDINE, which is present in such large quantities relative to the H^+ load that very little change in ICF $[H^+]$ occurs during a rise or fall in the ECF $[H^+]$. Hence, the concentration of buffer anion (buffer capacity) is more important than the change in $[H^+]$ in ICF buffering (page 13).

HCO_3 is also an important ICF buffer in metabolic acidosis. The quantity of HCO_3 in the ICF is close to that in the ECF because the ICF volume exceeds that of the ECF by 2-fold, but the ICF $[HCO_3]$ is only half that in the ECF. During metabolic acidosis, although there is only a small change in ICF $[H^+]$, the pulmonary response to lower the P_{CO_2} produces a corresponding change in ICF $[HCO_3^-]$—i.e., if the P_{CO_2} is halved at a constant $[H^+]$, the $[HCO_3^-]$ must also halve (see Equation 2).

Finally, phosphate could be an effective ICF buffer when only the ICF $[H^+]$ and its pK are under consideration; however, since the quantity of *inorganic* phosphate in the ICF is very low, it cannot buffer many H^+.

PHOSPHATE: THE URINE BUFFER, AND TITRATABLE ACID EXCRETION

The major urine buffer is phosphate (pK 6.8); it is effective as a buffer because the $[HPO_4^=]$ is relatively high in the urine, and the urine $[H^+]$ can be raised high enough so that it exceeds its pK. The H^+ excreted in the urine buffered with phosphate is referred to as *titratable acid* (TA). The filtered load of phosphate is not regulated by the need to increase renal H^+ excretion; therefore, the urine H_2PO_4 excretion has only a limited capacity (30–50 mmol/day) to increase renal H^+ excretion in times of severe acidemia. That role falls to urine ammonium excretion (page 23).

DYNAMICS OF ACID-BASE BALANCE

Daily H^+ Load

An individual eating the usual North American diet is faced with a net daily acid load of approximately 1 mmol/

BICARBONATE BUFFER SYSTEM

In biochemical terms, HCO_3^- is a relatively poor buffer. However, the presence of carbonic anhydrase, converting H_2CO_3 largely to CO_2, the solubility of CO_2, and especially the removal of CO_2 by the lungs all make the HCO_3^- buffer system more effective. Finally, the fact that the kidney synthesizes new HCO_3^- or excretes excess HCO_3^- completes this physiologic buffer system.

If the lungs were not present and a patient had a H^+ load that reduced the ECF [HCO_3^-] from 25 to 12.5 mM, the P_{CO_2} would be increased to 455 mm Hg (see Table below). The resultant [H^+] would be incompatible with life. However, because ventilation removes CO_2, and ventilation is stimulated by acidemia (Chapter 3, page 44), patients with a [HCO_3^-] of 12.5 mM should have a P_{CO_2} of 27 mm Hg; this produces a [H^+] of 52 nM. Had the P_{CO_2} been lowered only to 40 mm Hg, the [H^+] would be considerably higher (77 vs 52 nM).

Impact of CO_2 Removal on [H^+] With Halving of the [HCO_3^-]

Condition	[H^+]	pH	P_{CO_2}	HCO_3^-
Closed System	871	6.06	455	12.5
No Hyperventilation	77	7.11	40	12.5
Hyperventilation	52	7.29	27	12.5

TABLE 1–2. Major Buffers and Their Characteristics

	pK	Compartment pH	Content (mmol)	Basis of Effectiveness	Weakness
ECF					
HCO_3^-	6.1	7.4	350	• CO_2 removal by the lungs • Quantity	• Distance of pH from pK • Dependence on lungs
ICF*					
Imidazole	Close to 7	7.0	600–700	• Exceedingly large quantity	• Changes charge on ICF proteins†
HCO_3	6.1	7.0	330	• Relatively large quantity	• Dependence on lungs for CO_2 removal
Urine					
Phosphate	6.8	5–7	40 mmol per day	• pK higher than urine pH	• Little capacity to increase excretion rate

*Although ICF creatine phosphate is not a buffer, during acidemia it is hydrolyzed, rendering it capable of H^+ binding.

†This change in charge on ICF proteins can have important effects on enzyme activities, transporters, and ICF volume.

kg/day, which comes from the liver, largely owing to metabolism of the protein constituents of the diet (see page 15 for details). The relative magnitude of this acid load must be appreciated—the normal $[H^+]$ of the ECF is 40 nM, and the ECF volume is 14 L; therefore, the load of 70 mmol of H^+ per day is almost a millionfold greater than the total ECF H^+ content. It is therefore understandable that much more than 99% of this H^+ load must be buffered rather than remaining *free* in the ECF plus the ICF.

Buffering of Daily H^+ Load

Sixty percent of the H^+ load enters the ICF, where it is buffered both by the intracellular amino acid histidine and ICF HCO_3 buffer systems. The remainder stays in the ECF, where almost all the H^+ are buffered by HCO_3^-, with a minute proportion remaining as free H^+. The buffering process represents a temporary disposal of the H^+; the H^+ load must ultimately be removed.

The lungs play a critical role in the ECF buffering process, removing the CO_2 generated after the combination of H^+ with HCO_3^- (see page 13 for discussion).

Question (Answer on page 15)

6. How much would the ECF $[H^+]$ increase when a load of 70 mmol of H^+ is retained in a 70 kg person? At what rate would the $[HCO_3^-]$ fall if this patient lacked renal new HCO_3^- generation?

Buffering of H_2CO_3. Should the lungs be unable to excrete all the CO_2 produced, the level of H_2CO_3 rises and a new steady state occurs. Buffering of this acid load is somewhat different because the bicarbonate buffer system is impaired. For simplicity, we shall assume buffering is normal on the histidine system but is ineffective on the ICF and ECF HCO_3 systems.

Renal New Bicarbonate Generation

The 70 mmol of H^+ derived from the diet were initially buffered. Thereafter, it is the role of the kidney to regenerate the HCO_3^- that was consumed in the buffering process. The acid-base impact of the dietary H^+ load is reversed when the kidney has generated 70 mmol of HCO_3^-. Should

TABLE 1–3. Dietary Acid-Base Impact*

Nutrient	Product	mmol H^+/day†
Reactions Generating H^+		
SULPHUR-CONTAINING AMINO ACIDS		
Cysteine/Cystine Methionine	H_2SO_4	70
CATIONIC AMINO ACIDS		
Lysine	HCl	140
Arginine	HCl	
Histidine (½)	HCl	
Reactions Removing H^+		
ANIONIC AMINO ACIDS		
Glutamate	HCO_3^-	−70
Aspartate	HCO_3^-	−40
NONPROTEIN ORGANIC ANIONS	HCO_3^-	−60
ORGANIC PHOSPHATES	$H_2PO_4^-$	−30

*The net dietary H^+ load is approximately 70 mmol/day.

†H^+ load is calculated on the basis of the ingestion of 100 gm of protein from beefsteak.

Answer

6. Given a H^+ load of 70 mmol, 60% (42 mmol) enters the ICF where they are buffered, leaving 28 mmol to be buffered by the ECF. In a 70 kg person, the ECF volume is 14 L; therefore, the 28 mmol are removed with HCO_3^-, reducing its concentration by 2 mM. Thus, the [HCO_3^-] falls from 24 to 22 mM, and if the respiratory response to the acidemia is appropriate, the P_{CO_2} falls to 38 mm Hg. The [H^+] is therefore $23.9 \times 38/22 = 42$ nM. Therefore, owing to the buffering process, a H^+ load of 70 million nmol results in only a tiny rise in the ECF [H^+] (from 40 to 42 nM). In this setting, the [HCO_3^-] falls by 2 mM. Therefore, in the patient lacking renal HCO_3^- generation, the ECF [HCO_3^-] would be expected to fall by 2 mM per day. If 70 mmols of H^+ were retained daily for 5 days, the ECF [HCO_3^-] would fall by 10 mM to 14 mM. Thus the [H^+] would be $23.9 \times 30/14 = 51$ nM. Therefore, you can appreciate that despite the effectiveness of the buffering process, its capacity is limited, and significant acidemia results from the dietary acid load in the absence of renal new HCO_3 generation.

this process of new HCO_3^- generation fail, the patient becomes progressively acidemic, owing to the continued net negative balance for HCO_3^- (for more details, see Answer 6, page13).

> THERE ARE TWO COMPONENTS TO RENAL HCO_3^- GENERATION:
> 1. REABSORPTION OF FILTERED HCO_3^-
>   - achieved primarily by proximal H^+ secretion.
> 2. SYNTHESIS OF NEW HCO_3^-
>   - achieved by NH_4^+ production and excretion.

Reabsorption (Recycling) of Filtered HCO_3^-

Proximal Renal H^+ Secretion. The bulk of renal H^+ secretion occurs in the proximal part of the nephron. The major role of proximal H^+ secretion is to reabsorb filtered HCO_3^-, a process that does not lead to any net HCO_3^- gain (Fig. 1–2A). This is really HCO_3^- "RECYCLING" rather than reabsorption in the true sense of the word, as there is neither HCO_3^- gain nor loss, and there is no acid-base impact of the process. At a normal GFR (180 L/day), 4500 mmol of HCO_3^- are filtered, and approximately 85% of this HCO_3^- is "reclaimed" directly by proximal tubule H^+ secretion.

H^+ secretion in this nephron segment requires Na reabsorption (Fig. 1–2B). Note the important roles played by the Na K ATPase and luminal carbonic anhydrase in proximal H^+ secretion. In addition to the titration of filtered HCO_3^-, secreted H^+ are also trapped as $H_2PO_4^-$. There is also NH_4^+ secretion in the proximal convoluted tubule (see page 22 for more details). The $[HCO_3^-]$ at the end of the proximal tubule is normally close to 10 mM.

In summary, the major function of proximal H^+ secretion is to recycle the filtered HCO_3^-, preventing its loss in the urine. Proximal H^+ secretion is characterized by a high capacity (approaches 4000 mmol/day) but is not capable of achieving a steep $[H^+]$ gradient between cells and lumen (Table 1–4).

Regulation of Proximal H^+ Secretion. Since the Na^+/H^+ antiporter has a very large capacity for H^+ secretion but cannot achieve steep $[H^+]$ gradients, the regulation of the Na/H^+ antiporter is easily appreciated.

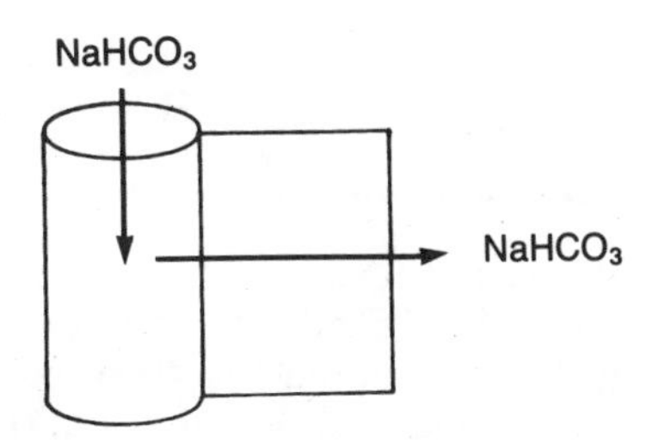

Figure 1–2A. Proximal H^+ secretion. The overall process: Filtered $NaHCO_3$ is recycled.

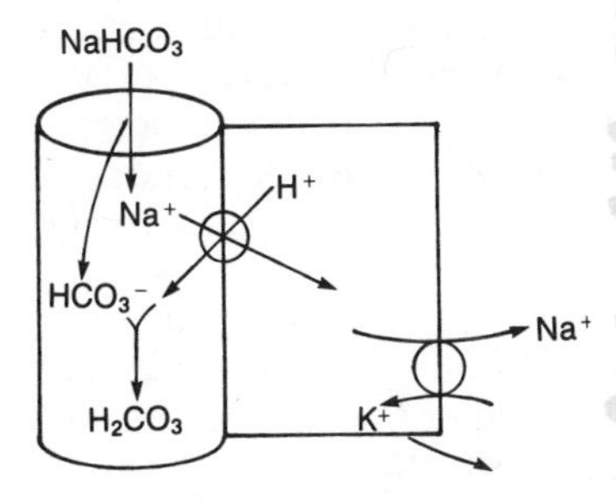

Figure 1–2B. Filtered Na^+ is reabsorbed, owing to a low ICF [Na] created by the Na K ATPase on the basolateral membrane. H^+ move from the ICF to the lumen on the Na^+/H^+ antiporter. To a minor extent, NH_4^+ can replace H^+ on this transporter (Fig. 1–4, page 25). The fate of this secreted H^+ is to react with filtered HCO_3^- to form luminal H_2CO_3. Thus, filtered HCO_3 disappears. This membrane is leaky; hence, steep $[H^+]$ gradients cannot be produced (see also Fig. 1–2C, page 19).

Regulation of proximal H^+ secretion:

1. Filtered load of HCO_3^-
2. Luminal $[H^+]$
3. Proximal ICF $[H^+]$
4. Avidity for Na reabsorption
5. Minor
   - —hypercalcemia
   - —low PTH levels
   - —hypokalemia

• **The Filtered Load of HCO_3.** When the plasma $[HCO_3]$ is 25 mM and the GFR 180 L/day, 4500 mmol of HCO_3 are filtered and 4000 reclaimed proximally. Lowering the plasma $[HCO_3^-]$ to 10 mM reduces its filtered load to 1800 mmol. In this case, H^+ secretion in the proximal convoluted tubule will have to be reduced by more than 50% (there are no other H^+ acceptors of quantitative importance).

• **Luminal $[H^+]$.** Should a patient be given a drug that inhibits luminal carbonic anhydrase, the luminal $[H^+]$ rises abruptly and brings H^+ secretion to a halt long before 85% of the filtered HCO_3 is reclaimed.

• **ICF $[H^+]$ in Proximal Tubular Cells.** A rise in the ICF $[H^+]$ stimulates H^+ secretion here for 2 reasons: a substrate effect of the ICF $[H^+]$ and the fact that the ICF $[H^+]$ activates the Na^+/H^+ antiporter. This activation is not very important during metabolic acidosis because of the small filtered load of HCO_3. Nevertheless, a modest increase in Na^+/H^+ antiporter activity can be seen during hypokalemia and with an elevated P_{CO_2}—conditions that may be associated with a higher ICF $[H^+]$. In contrast, when a $NaHCO_3$ load is administered, there is a fall in the ICF $[H^+]$. This leads to a lower rate of flux through the Na/H^+ antiporter and the prompt excretion of the excess HCO_3.

• **Avidity for Na^+ Reabsorption.** With a $NaHCO_3$ load, the increased Na load via ECF volume expansion seems to depress net $NaHCO_3$ recycling, contributing to the failure of the augmented filtered load of HCO_3 to increase H^+ secretion. In contrast, with a contracted ECF volume, proximal Na^+ reabsorption and thereby H^+ secretion tend to rise.

• **Minor Factors.** Hypercalcemia and low parathyroid hormone levels tend to augment proximal H^+ secretion.

TABLE 1–4. Net Renal H^+ Secretion

	Proximal	**Loop**	**Distal**
H^+ Pumped (mmol/d)	3900	400	270
H^+ Gradient (cell to lumen)	10:1	10:1	500:1
H^+ Secretory Process	Na^+/H^+	Na^+/H^+	H^+ ATPase
Luminal Carbonic Anhydrase	Yes	Yes ?	No

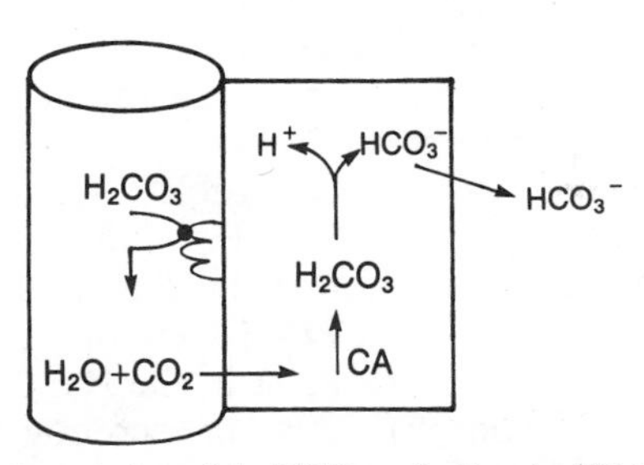

Figure 1–2C. Completion of the HCO_3 cycle. Luminal H_2CO_3 is converted to CO_2 + H_2O by a luminal carbonic anhydrase. The CO_2 + H_2O enter the cell and are converted to H^+ for secretion and HCO_3 for exit to the body by an intracellular carbonic anhydrase. Combining Figures 1–2B and 1–2C, the recycling of filtered $NaHCO_3$ is demonstrated.

The failure of proximal renal H^+ secretion results in

1. A fall in ECF [HCO_3^-], owing to renal HCO_3 excretion.
2. Inability to increase ECF [HCO_3^-] with HCO_3^- administration (low proximal threshold), i.e., HCO_3^- excretion rather than retention.
3. Alkaline urine (pH >7.0) in the presence of acidemia.

Proximal Renal Tubular Acidosis (pRTA). The failure of proximal H^+ secretion is called proximal renal tubular acidosis. The "apparent renal HCO_3^- threshold" represents that maximum ECF [HCO_3^-] at which proximal H^+ secretion is sufficient to maintain HCO_3^- reabsorption. In truth, there is no real threshold, as factors such as ECF volume expansion, which limits Na reabsorption, and a fall in ICF [H^+] diminish proximal H^+ secretion.

Question (Answer on page 21)

7. If the patient with pRTA develops diarrhea and the ECF [HCO_3^-] drops 2 mM below its usual value, how will the urine pH and [HCO_3^-] change?

Loop of Henle H^+ Secretion. Since close to 15% of filtered HCO_3 (675 mmol) leaves the proximal tubule and approximately 250 mmol enters the distal convoluted tubule, more than 400 mmol were removed either in the pars recta of the proximal tubule or in the thick ascending limb of the loop of Henle. Although less well characterized, H^+ secretion appears to mediate this process (Table 1–4, page 19).

Distal Nephron H^+ Secretion (Fig. 1–3). Distal nephron H^+ secretion occurs primarily in the cortical and medullary collecting duct and is an ACTIVE secretory process carried out by a H^+ ATPase (in contrast to the Na^+/H^+ antiporter of the proximal nephron). This H^+ secretion reclaims any remaining filtered HCO_3^- and therefore permits an increased renal NH_4^+ excretion, which results in generation of new HCO_3^-. However, in the absence of a H^+ acceptor, the free [H^+] in the tubular fluid rises quickly (but is only 0.01 mM), and H^+ secretion by the collecting duct ceases, owing to a gradient limit for H^+ secretion having been reached (the urine pH cannot be lowered below 4.0–4.5). Therefore, to ensure continued collecting duct H^+ secretion, NH_3 must be available to the collecting duct as a H^+ trap. Thus, a major regulator of collecting duct H^+ secretion is the available NH_3 (page 23).

Answer

7. Under some circumstances (e.g., after gastric HCl secretion), the patient with pRTA has ALKALINE urine because the FILTERED load of HCO_3^- exceeds the reduced capacity of the proximal H^+ secretory process to reclaim it. Thus, HCO_3^- is delivered distally, at which point some (but not all) is reclaimed, since distal H^+ secretion is a low-capacity system. The rest of the nonreclaimed HCO_3^- is excreted, rendering the urine alkaline. If some process (e.g., GI $NaHCO_3$ loss) lowers the filtered load of HCO_3^- so that it does not exceed the proximal H^+ secretory capacity, most of the HCO_3^- is then reabsorbed proximally (some HCO_3^- always leaves the proximal tubule, since the membrane is LEAKY, and steep $[H^+]$ gradients cannot be generated by the proximal nephron; i.e., the pH of the luminal fluid at the end of the proximal convoluted tubule is in the mid 6 range). Under these circumstances, a small enough amount of HCO_3^- is delivered distally so that distal H^+ secretion is sufficient to reabsorb this HCO_3^-, rendering the urine persistently HCO_3^--free. The remaining H^+ secretion may lower the urine pH to well below 6; how low the urine pH falls is determined by the quantity of NH_3 available to react with secreted H^+.

The features of distal H^+ secretion are:
- H^+ ATPase
- Less permeable membrane to H^+, ∴ steep $[H^+]$ gradient
- Absence of luminal carbonic anhydrase
- HCO_3^- exit step

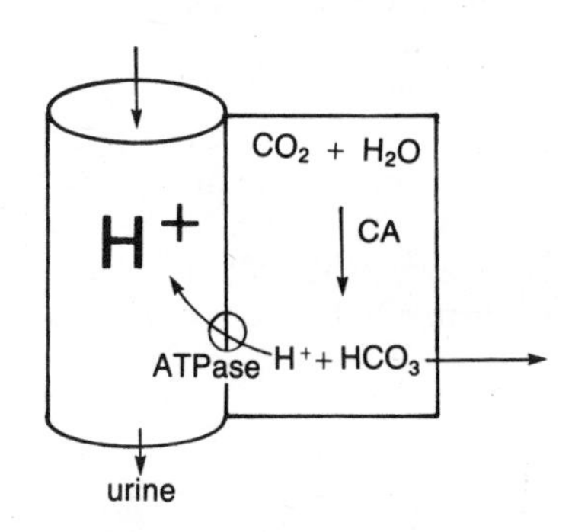

Figure 1–3. Distal nephron H^+ secretion. In the collecting duct, carbonic anhydrase–rich cells produce H^+ in the ICF. These are pumped into the lumen by an ATPase. In contrast to the proximal nephron, steep $[H^+]$ gradients are generated between the ICF and the lumen.

SYNTHESIS OF NEW HCO_3^-

Recognize by urine net acid excretion rate
- Add:
  - Urine NH_4^+
  - Urine $H_2PO_4^=$
  - Urine-free H^+ (trivial)
- Subtract:
  - Urine HCO_3

Urine NH_4^+ Excretion. The simple view of the acid-base role of renal NH_4^+ excretion is that glutamine metabolism leads to the formation of NH_3, which, through a series of complex events, becomes available to the collecting duct, where it can trap the secreted H^+ as NH_4^+ (HCO_3 is formed and added to the body when NH_4^+ is excreted). If secreted H^+ were not trapped as NH_4^+, the luminal $[H^+]$ would rise appreciably and prevent further collecting duct H^+ secretion.

One can equate renal NH_4^+ excretion with HCO_3^- generation (Fig. 1–5, page 25), and from a practical point of view, this serves the needs of the clinician. As presently perceived, the details of the sequence of events of glutamine metabolism and renal HCO_3^- generation are given on page 23.

Renal bicarbonate generation can fail for several reasons:

1. FAILURE OF NET HYDROGEN ION SECRETION
   - A. Damage to H^+ pump
   - B. Failure of pump mechanism
   - C. Back diffusion of secreted H^+
2. LACK OF NH_3 AVAILABILITY IN THE COLLECTING DUCT
   - A. Failure of glutamine metabolism
   - B. Failure to raise the $[NH_3]$ in the renal medulla

Whatever the basis, the failure of renal HCO_3^- generation is known as DISTAL RENAL TUBULAR ACIDOSIS (dRTA). As one can appreciate, there are many possible pathogenic mechanisms to explain the presence of dRTA; the specific basis must be established in each individual case. However, they all are characterized by a *relatively low rate of urine NH_4^+ excretion*.

RENAL NH_4^+ EXCRETION: THE "TRUE" STORY

NH_4^+ Production. Glutamine is metabolized in cells producing $2NH_4^+$ and the α-ketoglutarate$^=$ (α $KG^=$) anion. In order to have glutamine metabolism and NH_4^+ excretion lead to HCO_3^- generation, 2 additional events must occur:

1. Metabolism of α $KG^=$ to a neutral end-product (glucose or CO_2) consumes $2H^+$ (equivalent to the generation of 2 HCO_3^-).

$$\alpha\ KG^= + 2H^+ \rightarrow ½\ \text{glucose} + 2\ CO_2$$

2. The excretion of NH_4^+ in the urine prevents NH_4^+ delivery to the liver where $2H^+$ are produced (consuming 2 HCO_3^-) in the process of urea synthesis.

Glutamine metabolism is augmented by chronic metabolic acidosis and hypokalemia; there is a lag period of several days.

Glutamine metabolism is an ATP-producing process and may be limited by each cell's ATP utilization rate. The major ATP utilization process is Na reabsorption. Therefore, with reduced Na reabsorption (low GFR), ammoniagenesis may be impaired (one basis for impaired NH_4^+ excretion in renal failure).

NH_4^+ Transfer to the Urine. NH_4^+ enters the proximal tubular lumen either by nonionic diffusion of NH_3 (Fig. 1–4) or by NH_4^+ secretion on the Na^+/H^+ antiporter. NH_4^+ appears at the end of the proximal tubular lumen in the final concentration predicted by nonionic diffusion. Transfer of NH_4^+ to the lumen has *not* resulted in HCO_3^- generation or reclaimed filtered HCO_3^-.

Sixty percent of the NH_4^+ and the HCO_3^- present in the fluid at the end of the proximal tubule disappears during transit through the loop of Henle (LOH) as a result of the following reactions:

$$\text{(a)}\ NH_4^+ \rightleftharpoons NH_3 + H^+$$
$$\text{(b)}\ H^+ + HCO_3^- \rightleftharpoons CO_2 + H_2O$$

One mechanism is that as the fluid descends in the LOH, H_2O is reabsorbed, increasing the $[HCO_3^-]$; the fluid becomes alkaline, leading to reactions (a) and (b). Another mechanism is NH_4^+ reabsorption and H^+ secretion in the loop of Henle. Whatever the mechanism, the end result is the disappearance of NH_4^+ and HCO_3^- from the luminal fluid and the liberation of CO_2 and NH_3 in the medullary interstitium (Fig. 1–6).

The CO_2 formed in the LOH enters the collecting duct (CD) cell and is converted to H^+ and HCO_3^- by intracellular carbonic anhydrase. The H^+ are secreted into the lumen by H^+ ATPase, where they are bound by luminal NH_3 (liberated in the medullary interstitium). Thus the NH_4^+ and HCO_3^-, which were present in the LOH fluid, have reappeared—the NH_4^+ remaining in the tubular fluid lumen and being excreted, the HCO_3^- entering the blood representing recycled filtered HCO_3^- (Fig. 1–6). The only new HCO_3^- comes from the metabolism of α $KG^=$.

Pulmonary Physiology

The major role of the lungs from the acid-base point of view is to maintain the P_{CO_2} constant by removing the 10 mmol of CO_2 produced per minute by normal metabolism. Failure to remove this CO_2 leads to the generation of H^+ (Equation 1) and ACIDEMIA (respiratory acidosis), whereas excessive CO_2 removal results in ALKALEMIA (respiratory alkalosis).

The 3 components of pulmonary function that regulate gas exchange are VENTILATION, DIFFUSION, and PERFUSION. Abnormalities of CO_2 homeostasis are generally related to derangements of ventilation, and therefore we shall focus on the physiology and control of ventilation.

CONTROL OF VENTILATION

There are 3 components of ventilation:
1. CENTRAL CONTROL
2. RECEPTORS
3. RESPIRATORY MUSCLES and their workload

The Central Control of Ventilation. There are 2 components of the central control of ventilation. The rhythmic nature of ventilation is controlled by several centers in the medulla and in the pons, often referred to as the *central respiratory centers*. In addition, there is a voluntary control of breathing, which is mediated via the CORTEX, which is capable of overriding the brain stem, within limits.

Receptors. There are both central and peripheral chemoreceptors that play an important role in the control of ventilation. The central chemoreceptors are located on the ventral surface of the medulla and respond primarily to increases in the $[H^+]$ of the CSF with a stimulation of ventilation, and ventilation is inhibited when the $[H^+]$ falls. An increase in P_{CO_2} also stimulates ventilation via these receptors, and this is thought to be mediated by a rise in the $[H^+]$.

The peripheral chemoreceptors, primarily found in the carotid bodies, are responsible for the ventilatory response to hypoxia. These receptors also play a minor role in the ventilatory response to increased P_{CO_2} and acidemia.

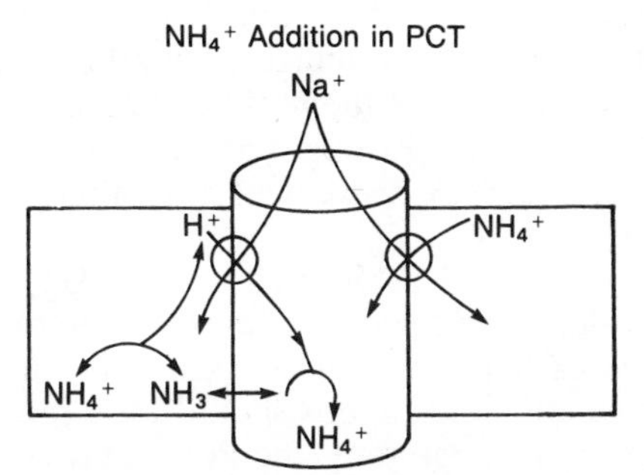

Figure 1–4. Mechanisms for NH_4 entry into the lumen of the proximal tubule. On the left, H^+ secretion on the Na/H^+ antiporter raises the luminal [H^+]. NH_3 is uncharged and enters the lumen, where it is converted to NH_4^+ (nonionic diffusion mechanism). On the right, NH_4^+ instead of H^+ is secreted on the Na/H^+ antiporter.

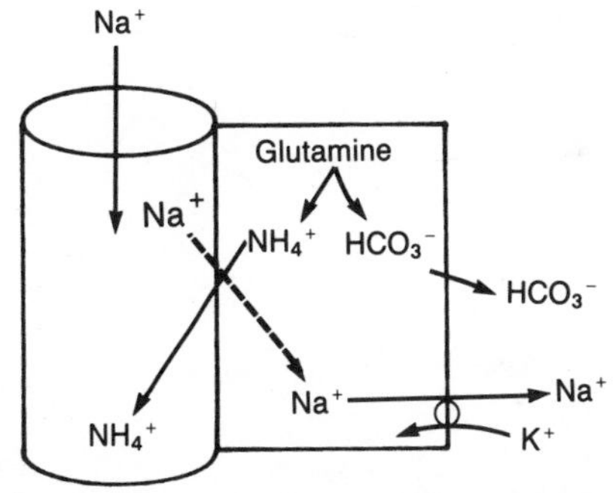

Figure 1–5. Mechanism of renal generation of new HCO_3. Glutamine metabolism leads to the formation of NH_4^+ and HCO_3^- (via α-ketoglutarate metabolism to CO_2). The NH_4^+ enters the tubular lumen (see Fig. 1–6); having no acid-base impact, the HCO_3^- enters the renal venous blood, representing *new HCO_3^- generation*.

Respiratory Muscles and Their Workload. The final determinants of ventilation, given the appropriate signals, are the respiratory muscles and the magnitude of the resistance that they must overcome. Muscle disease or metabolic disorders such as hypokalemia and hypophosphatemia and severe catabolic states may interfere with the actual function of the respiratory muscles themselves. In addition, changes in either the elastic load (lung and chest wall compliance) or resistive load (airway resistance) have a major impact on ventilation.

Whereas the normal P_{CO_2} is 38–42 mm Hg, in states of metabolic acidosis and hypoxia one expects to see an appropriate degree of hyperventilation. Similarly, in association with metabolic alkalosis, there should be a modest degree of hypoventilation. Therefore, the P_{CO_2} must be assessed in the perspective of the pathophysiologic state of the patient, considering the existing stimulators and suppressors of ventilation.

Abnormalities of Acid-Base Balance

The pathophysiology of the 4 primary acid-base disorders are discussed in the chapter dealing with each specific disorder.

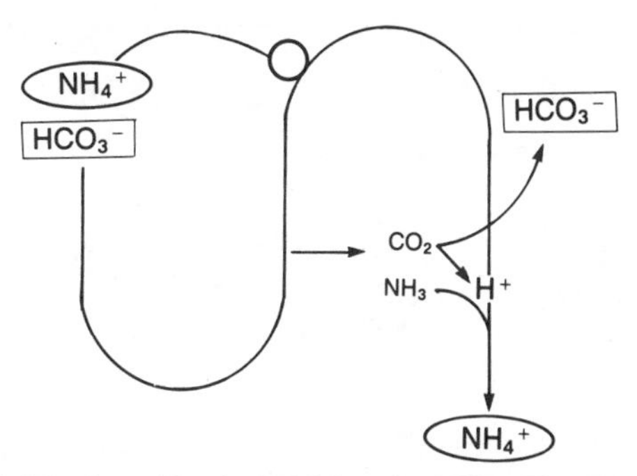

Figure 1–6. The fate of luminal HCO_3^- plus NH_4^+. The filtered HCO_3 is removed when it is converted to CO_2 in the loop of Henle. This CO_2 enters collecting duct cells, resynthesizing HCO_3, which returns to the body. The net result is a HCO_3 recycling. Luminal NH_4^+ becomes NH_3 in the medulla and then luminal NH_4^+ again in the collecting duct (see text for details).

THE PRODUCTION OF ACIDEMIA BY AN ELEVATION IN CO_2

$$CO_2 + H_2O \overset{CA}{\rightleftharpoons} H_2CO_3 \rightleftharpoons H^+ + HCO_3^-$$

Accumulation of CO_2 displaces the HCO_3^- buffer system equilibrium to the right, generating *EQUIMOLAR* amounts of HCO_3^- and H^+. However, H^+ is present only in NANOMOLAR quantities, whereas HCO_3^- is present in MILLIMOLAR quantities. Therefore, even though the vast majority of H^+ generated by CO_2 accumulation are buffered in the ICF owing to a rise in ICF [H^+], there is also an INCREASE in ECF [H^+] and ACIDEMIA, despite the coincident increase in [HCO_3^-].

Suggested Reading

Aronson PS: Mechanisms of active H^+ secretion in the proximal tubule. Am J Physiol 245:647–659, 1983.

Halperin ML, Goldstein MB, Stinebaugh BJ, et al: Biochemistry and physiology of ammonium excretion. In Seldin DW and Giebisch G (eds): The Kidney. Physiology and Pathophysiology. New York, Raven Press, 1985, pp 1471–1490.

Halperin ML and Jungas RL: Metabolic production and renal disposal of hydrogen ions. Kidney Int 24:709–713, 1983.

Ives HE and Rector FC: Proton transport and cell function. J Clin Invest 73:285–290, 1984.

Schwartz WB and Relman AS: A critique of the parameters used in the evaluation of acid–base disorders. N Engl J Med 268:1382–1388, 1963.

Swan RC, Pitts RF, and Madisso H: Neutralization of infused acid by nephrectomized dogs. J Clin Invest 34:205–212, 1955.

2

THE CLINICAL APPROACH TO ACID-BASE DISORDERS

THE BIG PICTURE

You need 4 parameters to analyze the patient's acid-base status; their normal values are:

Plasma $[H^+]$**:** **40 ± 2 nM (pH: 7.40 ± 0.02)**

Plasma P_{CO_2}**:** **40 ± 3 mm Hg**

Plasma $[HCO_3^-]$**:** **24 ± 2 mM**

Plasma anion gap: **12 ± 2 mEq/L**

The Henderson equation relates 3 of these parameters; you can calculate the third knowing any 2.

$$[H^+] = 23.9 \times P_{CO_2}/[HCO_3^-]$$

There are 4 primary acid-base disorders:

Metabolic Acidosis: **Increased** $[H^+]$ **and reduced** $[HCO_3]$

Metabolic Alkalosis: **Reduced** $[H^+]$ **and increased** $[HCO_3]$

Respiratory Acidosis**: Increased** $[H^+]$ **and** P_{CO_2}

Respiratory Alkalosis**: Reduced** $[H^+]$ **and** P_{CO_2}

Specific physiologic responses occur as the result of acid-base disorders, which return the plasma $[H^+]$ **toward normal, but may only return the plasma** $[H^+]$ **to normal in chronic respiratory alkalosis.**

In metabolic acidosis, compare the increase in plasma anion gap with the fall in plasma $[HCO_3^-]$**; expect a 1:1 relationship.**

The clinical and laboratory data must be evaluated in concert to reach the final acid-base diagnosis.

Several acid-base disorders may coexist.

pH vs $[H^+]$

The disadvantages of the pH relative to the $[H^+]$ have been discussed in Chapter 1. There are several "bedside" ways to establish the $[H^+]$ for any given pH. Two are illustrated and a reference table is provided below.

pH	$[H^+]$
7.40	40
7.38	42
7.42	38

1. There is a near-linear relationship between the $[H^+]$ and the pH in the range of pH 7.28 to 7.48. At pH 7.40, the $[H^+]$ is 40 nM (note that if you drop the 7 and the decimal point you have the same number in both columns). Then, beginning from 40, if the pH falls by 0.02 units, the $[H^+]$ rises by 2 nM and vice versa. In the same way, if the pH is 7.30, you should calculate that the $[H^+]$ is 50 nM.

pH	$[H^+]$
6.90	125
7.00	100
7.10	80
7.20	64

2. If one pH and the corresponding $[H^+]$ are known, all subsequent values can be easily calculated from the known relationship that a 0.1 unit increase in pH $= 0.8 \times [H^+]$. Given that pH 7.00 $= [H^+]$ of 100 nM, a rise in pH of 0.1 (pH 7.10) $= 0.8 \times 100$ or a $[H^+]$ of 80 nM. For values less than 7.00, divide by 0.8. Intermediate values are calculated by interpolation, e.g., pH 7.03 is $100 - 0.3 \times (100 - 80) = 94$ nM.

Log Table for Interconversion of pH and $[H^+]$*

pH	$[H^+]$	pH	$[H^+]$	pH	$[H^+]$	pH	$[H^+]$
.01	9772	.26	5495	.51	3090	.76	1738
.02	9550	.27	5370	.52	3020	.77	1698
.03	9333	.28	5248	.53	2951	.78	1660
.04	9120	.29	5129	.54	2884	.79	1622
.05	8913	.30	5012	.55	2818	.80	1585
.06	8710	.31	4898	.56	2754	.81	1549
.07	8511	.32	4786	.57	2692	.82	1514
.08	8318	.33	4677	.58	2630	.83	1479
.09	8128	.34	4571	.59	2570	.84	1445
.10	7943	.35	4467	.60	2512	.85	1413
.11	7762	.36	4365	.61	2455	.86	1380
.12	7586	.37	4266	.62	2399	.87	1349
.13	7413	.38	4169	.63	2344	.88	1318
.14	7244	.39	4074	.64	2291	.89	1288
.15	7079	.40	3981	.65	2239	.90	1259
.16	6918	.41	3890	.66	2188	.91	1230
.17	6761	.42	3802	.67	2138	.92	1202
.18	6607	.43	3715	.68	2089	.93	1175
.19	6457	.44	3631	.69	2042	.94	1148
.20	6310	.45	3548	.70	1995	.95	1122
.21	6166	.46	3467	.71	1950	.96	1096
.22	6026	.47	3388	.72	1905	.97	1072
.23	5888	.48	3311	.73	1862	.98	1047
.24	5754	.49	3236	.74	1820	.99	1023
.25	5623	.50	3162	.75	1778	.00	1000

*For example: If pH = 6.99, then $[H^+] = 102.3$; if pH = 7.01, then $[H^+] = 97.7$.

ILLUSTRATIVE CASE (Answer on page 31)

1. A 20 year old diabetic developed abdominal pain and diarrhea. Because of the abdominal pain, he reduced his food intake and also reduced his insulin dose. As the diarrhea persisted, he saw his physician, and the following results were obtained.

Na	130 mM	$[H^+]$	60 nM	(pH = 7.22)
K	5.0	P_{CO_2}	25 mm Hg	
Cl	100	HCO_3	10 mM	

Is this a simple or a mixed acid-base disturbance?

HOW TO IDENTIFY ACID-BASE DISORDERS

When you are faced with a set of blood gas and electrolyte results, the approach outlined in the flow chart on page 31 should lead to a correct analysis. We prefer to think in terms of the $[H^+]$, and we have provided several means to convert pH to $[H^+]$ on page 29. For those who prefer pH, the principles are the same: a low $[H^+]$ is a high pH and vice versa.

We use the plasma $[H^+]$, $[HCO_3^-]$, P_{CO_2}, and anion gap initially. The P_{O_2} has no direct application with respect to the acid-base analysis; however, calculating the alveolar-arterial (A-a) O_2 gradient provides additional important information (see Chapter 5, page 126 for discussion).

Acid-Base Parameters and Their Normal Values
Plasma $[H^+]$ = 40 ± 2 nM (pH 7.40 ± 0.02)
Plasma $[HCO_3^-]$ = 24 ± 2 mM
Plasma P_{CO_2} = 40 ± 3 mm Hg
Plasma anion gap = 12 ± 2 mEq/L

One should begin with the $[H^+]$; if it is increased, the patient has acidemia, and there are 2 potential causes: metabolic and/or respiratory acidosis.

Metabolic Acidosis. This is characterized by a low $[HCO_3^-]$ and high $[H^+]$; the expected compensatory response is lowering of the blood P_{CO_2} by hyperventilation.

Respiratory Acidosis. This is characterized by an INCREASED P_{CO_2}, and the expected compensatory response is to increase the $[HCO_3^-]$. This compensation is minimal

Answer

1. The patient has a high plasma $[H^+]$ and low $[HCO_3^-]$; therefore, he has metabolic acidosis. The plasma anion gap is wide (20 mEq/L); therefore, the patient has a metabolic acidosis of the wide anion gap variety. However, the increase in plasma anion gap is 8 mEq/L (20 − 12), and the $[HCO_3^-]$ has fallen by 15 mM (25 − 10 mM); therefore, the fall in $[HCO_3^-]$ is greater than the increase in plasma anion gap, indicating that a component of the metabolic acidosis is of the $NaHCO_3$ loss variety. The fall in P_{CO_2} from 40 mm Hg (15 mm Hg) is appropriate for the fall in the plasma $[HCO_3^-]$ from 25 mM to 10 mM (25 − 10 = 15 mM). Thus, this patient has metabolic acidosis, part of which is of the wide AG variety (confirmed as ketoacidosis) and part of which is due to $NaHCO_3$ loss (diarrhea).

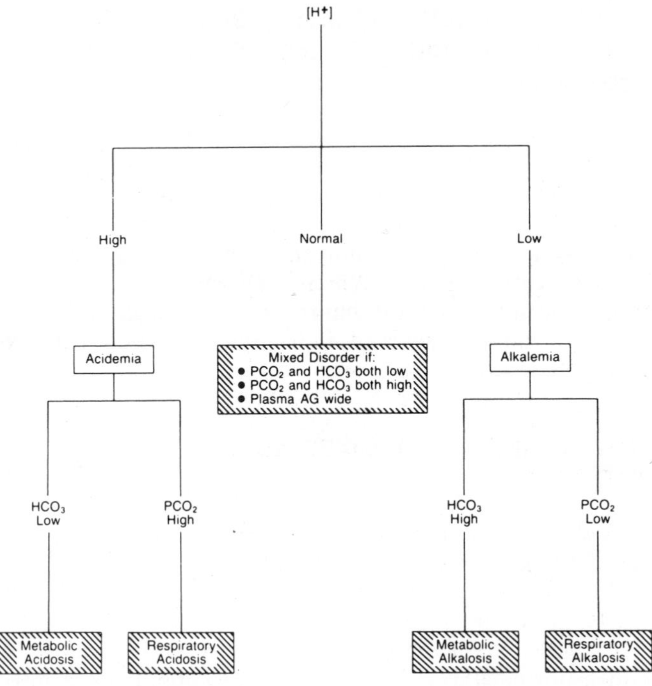

Flow Chart 2–1. Initial diagnosis of acid-base disorders. Start with the plasma $[H^+]$. The information to be interpreted to make a final diagnosis is indicated in open boxes; the final diagnoses are shown in the shaded boxes (see text for details). Plasma AG = plasma unmeasured anion gap.

in acute respiratory acidosis and larger in chronic respiratory acidosis.

If the $[H^+]$ is low, the patient has an alkalemia, and again there are 2 potential causes:

Metabolic Alkalosis. This is characterized by an increase in plasma $[HCO_3^-]$ and a fall in $[H^+]$. The expected compensatory response is hypoventilation ($\uparrow$ PCO_2), which is usually modest, owing to respiratory stimulation from the resultant hypoxia. However, if the patient is receiving O_2, and hypoxia is thus prevented, hypoventilation can be more profound.

Respiratory Alkalosis. This is characterized by a low PCO_2 and a low $[H^+]$. The expected compensation is a reduction in plasma $[HCO_3^-]$, which, like respiratory acidosis, is modest in acute disorders and more significant in chronic disorders.

EXPECTED COMPENSATORY RESPONSE TO THE PRIMARY ACID-BASE DISORDERS (Table 2–1)

It has been empirically observed that when a patient has 1 of the above 4 primary acid-base disturbances, a predictable compensatory response occurs to return the $[H^+]$ toward normal. Only in chronic respiratory alkalosis does the $[H^+]$ actually return within the normal range with the compensatory response. When faced with an acid-base disorder, one should ensure that the appropriate compensation has occurred, its absence indicating a second acid-base disturbance.

HOW TO RECOGNIZE MIXED ACID-BASE DISORDERS

Acid-base disorders may present as 2 or 3 coexisting disorders, e.g., alcoholic ketoacidosis may coexist with metabolic alkalosis (from vomiting) and respiratory alkalosis (due to aspiration pneumonitis).

It is possible for the patient to have an acid-base disorder with a normal $[H^+]$. In this case, the patient has a mixed disorder (alkalosis and acidosis resulting in a normal $[H^+]$. In fact, the PCO_2 and $[HCO_3^-]$ may also be normal, and the

TABLE 2–1. Expected Compensatory Responses in Primary Acid-Base Disorders

Disorder	Response
Metabolic Acidosis	• For every mM fall in plasma [HCO_3^-] from 25 mM, the P_{CO_2} should fall by 1 mm Hg from 40 mm Hg
Metabolic Alkalosis	• For every mM rise in plasma [HCO_3^-] from 25 mM, the P_{CO_2} should rise by 0.7 mm Hg from 40 mm Hg
Respiratory Acidosis	
Acute	• For every mm Hg rise in P_{CO_2} from 40, the plasma [H^+] should rise by 0.77 nM from 40 nM or the [HCO_3^-] should rise by 0.1 mM from 25 mM • Alternatively, for every 2-fold increase in P_{CO_2}, the plasma [HCO_3^-] increases by 2.5 mM
Chronic	• For every mm Hg rise in P_{CO_2} from 40, the [H^+] should rise by 0.32 nM from 40 nM or the plasma [HCO_3^-] should rise by 0.3 mM from 25 mM
Respiratory Alkalosis	
Acute	• For every mm Hg P_{CO_2} fall from 40, the plasma [H^+] should fall by 0.74 nM from 40 nM • Alternatively, for every 2-fold decrease in P_{CO_2}, the plasma [HCO_3^-] decreases by 2.5 mM
Chronic	• For every mm Hg fall in P_{CO_2} from 40, the plasma [H^+] should fall by 0.17 nM or [HCO_3^-] by 0.5 mM

only clue to the acid-base disorder is the increased plasma anion gap.

Some have approached the identification of mixed acid-base disorders by the use of nomograms; however, we have not adopted that approach because without the nomogram in your possession, you may have great difficulty. We prefer to provide the facts upon which the nomograms are based. Applying simple guidelines permits you to interpret the laboratory values correctly.

Questions (Answers on page 35)

2. A 23 year old woman with rheumatoid arthritis increased her dose of salicylates because of a flare-up. She then developed epigastric pain and vomiting that persisted for 2 days. She went to the local hospital where the following blood results were obtained:

$[H^+]$ = 18 nM (pH = 7.72) P_{CO_2} = 25 mm Hg

What is (are) your acid-base diagnosis(es)?

3. A 50 year old woman underwent intestinal bypass for morbid obesity. As she was having 10–15 watery stools per day, she was treated with tincture of opium and was found somnolent and hypotensive in the morning. Plasma values were

Na	130 mM	$[H^+]$	96 nM (pH 7.02)
K	3.2 mM	P_{CO_2}	40 mm Hg
Cl	102 mM	HCO_3	10 mM

A. What is (are) her acid-base diagnosis(es)?
B. What treatment would you consider?

GUIDELINES FOR THE DIAGNOSIS OF MIXED DISORDERS

1. Calculate the plasma anion gap—if it is increased by >5 mEq/L, the patient most likely has METABOLIC ACIDOSIS (see Chapter 3, page 48 for details).

2. Compare the magnitude of the fall in plasma $[HCO_3^-]$ with the increase in plasma anion gap; these changes should be similar in magnitude. If there is a gross discrepancy (>5 mEq/L), there is a mixed disturbance present.

Answers

2. The woman is very alkalemic ($[H^+]$ = 18 nM, pH = 7.72). Her $[HCO_3^-]$ can be calculated from the Henderson equation:

$$[HCO_3^-] = 23.9 \times Pco_2/[H^+] = 33 \text{ mM}$$

Thus, she has METABOLIC ALKALOSIS (low $[H^+]$ and elevated plasma $[HCO_3^-]$). However, her Pco_2 is low (the expected compensation for metabolic alkalosis is to hypoventilate to return the $[H^+]$ towards normal), and she is HYPERVENTILATING, therefore she has a secondary primary acid-base disorder—RESPIRATORY ALKALOSIS—and this is why she is so alkalemic. Her metabolic alkalosis was secondary to vomiting and HCl loss (Fig. 4–1), and her respiratory alkalosis was secondary to salicylate intoxication.

3A. The patient is very acidemic ($[H^+]$ = 96 nM, pH = 7.02). The $[HCO_3^-]$ is low (10 mM); therefore, she has a metabolic acidosis. The plasma anion gap is wide (18 mEq/L), i.e., increased by about 6 mEq/L; however, the $[HCO_3^-]$ has fallen by 15 mM (from 25 mM). Thus the fall in $[HCO_3^-]$ exceeds the increase in the plasma anion gap, indicating 2 components to the metabolic acidosis; part due to the accumulation of an organic acid, as reflected by the increase in the plasma anion gap (presumably lactic acidosis) and part due to $NaHCO_3$ loss in the diarrhea. The patient's Pco_2 is 40 mm Hg, which is *not* what you should expect during metabolic acidosis with a plasma $[HCO_3^-]$ of 10 mM—such a patient should hyperventilate to a Pco_2 of $40 - 15 = 25$ mm Hg as a compensatory response. Thus, this patient also has a respiratory acidosis. Therefore, this patient has 3 acid-base disturbances:

1. Lactic acidosis (hypotension and hypoxia)
2. Metabolic acidosis due to $NaHCO_3$ loss
3. Respiratory acidosis

3B. The treatment is determined by the underlying causes for the acid-base disturbances. Presumably the lactic acidosis and the respiratory acidosis are both due to the central nervous system suppression by the narcotic. Therefore, treatment with Narcan would be an appropriate first step. One could also stimulate the patient (verbally and physically) to breathe. If the patient did not respond to the Narcan or if her condition deteriorates, mechanical ventilation would give the quickest control of the acidemia. If one reduced her Pco_2 to 25 mm Hg, her plasma $[H^+]$ would be $23.9 \times 25/10 = 60$ nM (pH = 7.22). The patient is very acidemic and has lost some $NaHCO_3$; therefore, one could also give $NaHCO_3$ to alleviate the severe acidemia.

If the plasma anion gap has risen LESS than the fall in plasma [HCO_3^-], this suggests that a component of the metabolic acidosis is of the $NaHCO_3$ loss type.

On the other hand, if the increase in plasma anion gap is much greater than the fall in [HCO_3^-], this suggests that there is a coexistent metabolic alkalosis (i.e., an additional source of HCO_3^-].

3. In METABOLIC ACIDOSIS, for every mM fall in [HCO_3^-] from 25 mM, the Pco_2 should fall by 1 mm Hg from 40. If the Pco_2 is significantly lower, the patient has coexistent RESPIRATORY ALKALOSIS; if higher, RESPIRATORY ACIDOSIS.

4. In METABOLIC ALKALOSIS, expect to see a 0.7 mm Hg INCREASE in Pco_2 for every mM INCREASE in plasma [HCO_3^-]. If that increase is significantly greater, the patient has coexistent RESPIRATORY ACIDOSIS; if absent, coexistent RESPIRATORY ALKALOSIS.

5. In RESPIRATORY acid-base disturbances, one must distinguish between ACUTE and CHRONIC (more than 3–4 days) conditions on CLINICAL GROUNDS, as different degrees of physiologic compensation are present.

6. In ACUTE RESPIRATORY ACIDOSIS, anticipate a small increase in [HCO_3^-]. For every mm Hg increase in Pco_2, expect a 0.77 nM increase in [H^+]. If the increase in [H^+] is much greater, there is a coexistent METABOLIC ACIDOSIS; if much smaller, a coexistent METABOLIC ALKALOSIS.

7. In CHRONIC RESPIRATORY ACIDOSIS, the increase in plasma [HCO_3^-] is greater, owing to altered renal H^+ secretion. For every mm Hg increase in Pco_2, expect a 0.32 nM increase in [H^+]. This translates to a 3 mM INCREASE in plasma [HCO_3^-] for every 10 mm Hg INCREASE in Pco_2. If the change in plasma [HCO_3^-] is significantly greater, the patient has coexistent METABOLIC ALKALOSIS; if significantly smaller, coexistent METABOLIC ACIDOSIS.

8. In ACUTE RESPIRATORY ALKALOSIS, there is a small decrease in plasma [HCO_3^-]. For every mm Hg decrease in Pco_2 from 40 mm Hg, expect a 0.74 nM decrease in [H^+] from 40 nM. If the decrease in [H^+] is significantly greater, there is a coexistent METABOLIC ALKALOSIS; if significantly less, there is a coexistent METABOLIC ACIDOSIS.

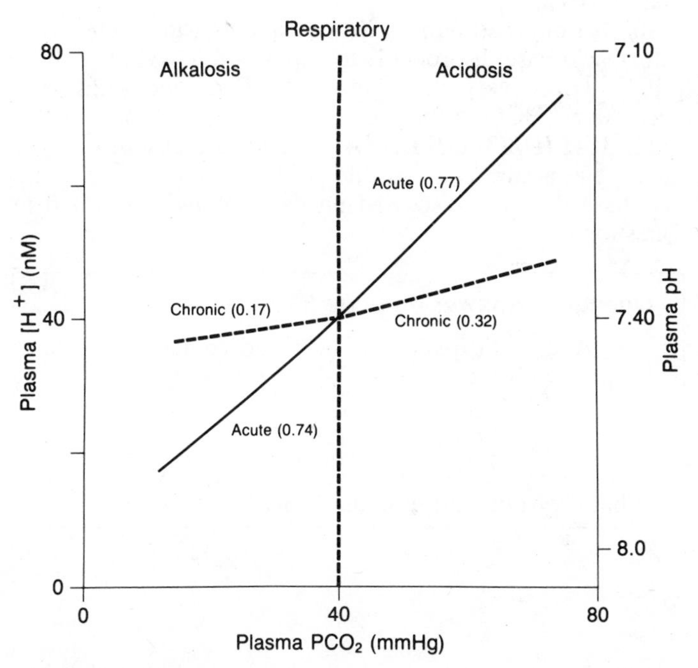

Figure 2–1. Expected compensatory responses in respiratory acid base disorders. The figure depicts the empirically derived near-linear relationships between plasma $[H^+]$ and P_{CO_2} in acute and chronic respiratory acidosis and alkalosis. A significant deviation from the expected value indicates the presence of another acid-base disturbance (see text for details).

9. In CHRONIC RESPIRATORY ALKALOSIS, there is a greater fall in plasma [HCO_3^-]; in fact, this is the only acid-base disturbance in which the physiologic compensatory changes may return the plasma [H^+] to the normal range. For every mm Hg decrease in PCO_2, expect a 0.17 nM decrease in [H^+]. In terms of plasma [HCO_3^-], for every 10 mm Hg decrease in PCO_2 from 40, expect a 5 mM fall in plasma [HCO_3^-].

10. An alternate approach for patients with acute respiratory acid-base disorders is to expect a 2.5 mM *increase* in [HCO_3^-] for every *doubling* of the PCO_2 and a 2.5 mM *decrease* in [HCO_3^-] for every *halving* of the PCO_2.

11. *VALIDITY OF DATA*: One should ensure that the data are as accurate as possible. The mechanisms to validate acid-base data are discussed on page 39 and illustrated in Question 4.

Question (Answer on page 39)

4. A diabetic patient is admitted comatose to the hospital with the following laboratory results:

Na	130 mM	[H^+]	80 nM (7.10)
K	5.0 mM	PCO_2	30 mm Hg
Cl	94 mM	HCO_3	13 mM

What are your acid-base diagnoses?

VALIDITY OF LABORATORY RESULTS

Clearly, a correct analysis of the laboratory results presumes that the data are accurate. There are 2 ways in which you can detect laboratory errors. The first is to calculate the plasma anion gap. If it is very low or negative, there is probably an error in one of the electrolyte values, unless the patient has multiple myeloma or one of the conditions listed in Chapter 3, page 51.

The second way to evaluate the laboratory results is to insert the $[H^+]$, P_{CO_2}, and $[HCO_3^-]$ (as reflected in the plasma electrolyte values) into the Henderson equation. If there is a substantial error ($>10\%$), one of the 3 parameters is incorrect; and if the discrepancy is great enough to change the diagnosis, the tests should be repeated and the error identified (see answer below).

Answer

4. Inserting the numbers in the Henderson equation:

$$
\begin{aligned}
[H^+] &= 23.9 \times P_{CO_2}/[HCO_3^-] \\
80 &= 23.9 \times 30/13 \\
80 &= 56
\end{aligned}
$$

Therefore, clearly, at least 1 of the 3 parameters is in error. When one is faced with these laboratory results, an appropriate approach would be as follows:

Whatever the scenario, the patient has a metabolic acidosis with an increased plasma anion gap, presumably diabetic ketoacidosis, and that diagnosis should be confirmed with the appropriate tests (serum ketones, blood glucose) and initial treatment instituted. One should also repeat the blood tests to clarify the basis of the discrepancy.

Suggested Reading

Adrogue HJ, Brensilver J, and Madias NE: Changes in the plasma anion gap during chronic metabolic acid-base disturbances. Am J Physiol 235:291–297, 1978.

Fagan TJ: Estimation of hydrogen ion concentration. N Engl J Med 288:915, 1973.

Gabow PA: Disorders associated with an altered anion gap. Kidney Int 27:472–483, 1984.

Kassirer JP and Bleich HL: Rapid estimation of plasma carbon dioxide tension from pH and total carbon dioxide content. N Engl J Med 272:1067, 1965.

Narins RG and Emmett M: Simple and mixed acid-base disorders: A practical approach. Medicine 59:161, 1980.

Oh MS and Carroll HJ: The anion gap. N Engl J Med 297:814–817, 1977.

3

METABOLIC ACIDOSIS

THE BIG PICTURE

Metabolic acidosis occurs with acid gain (other than H_2CO_3) or $NaHCO_3$ loss. Both result in a rise in [H^+] and fall in the plasma [HCO_3^-].

Metabolic acidosis often occurs as a complication of catastrophic illness (shock, sepsis), adding to the seriousness of the clinical setting.

Metabolic acidosis is NOT a specific diagnosis, and the underlying cause must be sought, as specific therapy may be life-saving (e.g., insulin for diabetic ketoacidosis, ethanol for methanol intoxication).

The rate of H^+ input may be high (organic acid overproduction) or low (metabolic acidosis due to renal failure). In the latter case, the plasma [HCO_3^-] de-

PHYSIOLOGIC CAPSULES

Acid-Base Homeostasis

BICARBONATE CONSUMPTION
- DIET H^+ LOAD (1 mmol/kg/day)
- EXCESSIVE ENDOGENOUS H^+ PRODUCTION (ketoacidosis, lactic acidosis)
- HCO_3^- LOSS (gastrointestinal, renal)

BUFFERING
- 60% intracellular (histidine, HCO_3^-)
- 40% extracellular (HCO_3^-)

NEW BICARBONATE GENERATION BY KIDNEYS
- GLUTAMINE METABOLISM
- NH_4^+ TRANSFER TO URINE

Bicarbonate Buffer System

$$H^+ + HCO_3^- \rightleftharpoons H_2CO_3 \rightleftharpoons H_2O + CO_2$$

The major ECF buffer
Effective because of CO_2 removal via lungs
Relationships: see the Henderson equation:

$$[H^+] = 23.9 \times Pco_2/[HCO_3^-]$$

Renal New Bicarbonate Generation

PREVENT HCO_3^- LOSS IN URINE (RECYCLE HCO_3^-)
- PROXIMAL H^+ SECRETION
  - 4000 mmol/day
  - Na^+/H^+ antiporter
  - Regulation of proximal Na^+/H^+ antiporter:
    - Filtered load of HCO_3^- most important
    - ICF $[H^+]$ (plasma $[H^+]$, Pco_2, $[K^+]$)
    - Avidity for Na^+ (ECF volume contraction)

NH_4^+ EXCRETION (NEW HCO_3^- GENERATION)
- Distal H^+ secretion
  - Active H^+ secretion (H^+ ATPase)
  - Steep $[H^+]$ gradient generated (lumen: cell)
  - Stimulated by aldosterone, acidemia, luminal $[NH_3]$
  - Up to 400 mmol NH_4^+ excreted/day
- NH_3 AVAILABILITY TO COLLECTING DUCT
  - Ammoniagenesis (impaired by hyperkalemia, low GFR)
  - Loop of Henle NH_3 generation (impaired in diseases affecting the medullary interstitium)

clines slowly but progressively; thus, the acidemia might be equally severe in both cases.

The plasma anion gap provides a useful clue to determine the basis of the metabolic acidosis.

The impact of metabolic acidosis upon the $[H^+]$ is minimized by a reduction in P_{CO_2}. For every 1 mM fall in $[HCO_3^-]$ *from 25 mM*, expect a 1 mm Hg fall in P_{CO_2} *from 40 mm Hg*.

Profound derangements in the plasma $[K^+]$ may accompany either metabolic acidosis or its therapy. At times, this may pose a greater threat to the patient than the acidemia itself.

The therapeutic role of $NaHCO_3$ in metabolic acidosis varies; for example, its use is important when the degree of acidemia is severe but may be dangerous in patients with metabolic acidosis who have severe hypokalemia. The decision to give $NaHCO_3$ or not must be made in each specific case.

ILLUSTRATIVE CASE

These plasma results were obtained in a patient presenting with weakness.

Na^+	140 mM	H^+	84 nM	(pH 7.07)
K^+	5.8 mM	P_{CO_2}	35 mm Hg	
Cl^-	115 mM	HCO_3^-	10 mM	

QUESTIONS (Answers on page 43)

1. What is (are) your acid-base diagnosis(es)?
2. What would the plasma $[H^+]$ be if the respiratory response were normal?
3. What are the likely causes for the metabolic acidosis in this case?
4. What is the significance of the plasma $[K^+]$?
5. What are the specific considerations with respect to HCO_3^- therapy in this case?
6. Must all patients with metabolic acidosis have a high plasma $[H^+]$?
7. Can a patient have a low plasma $[HCO_3^-]$ and not have metabolic acidosis?
8. Can one have a persistently alkaline urine (i.e., no HCO_3^- generation) and maintain acid-base balance?

Answers

1. The patient has a MIXED acid-base disorder: METABOLIC ACIDOSIS ($[HCO_3^-]$ = 10 mM with acidemia) and RESPIRATORY ACIDOSIS (during acidemia, a $[HCO_3^-]$ of 10 mM should be accompanied by a P_{CO_2} of 25 mm Hg, not 35 mm Hg).

2. If the P_{CO_2} were the expected 25 mm Hg, the $[H^+]$ would be 63 nM instead of 84 nM (a blood pH of 7.20 instead of 7.07).

3. The patient has a near normal plasma anion gap and therefore has lost $NaHCO_3$. You should suspect initially either reduced renal new HCO_3^- generation (a renal acidification defect) or $NaHCO_3$ loss via the gastrointestinal tract or the urinary tract (see Table 3–1, page 45, for a more complete list).

4. Hyperkalemia is an important diagnostic clue. Its presence suggests that the basis of the metabolic acidosis is most likely a reduced rate of renal NH_4^+ excretion. Hyperkalemia is unusual among the other causes of metabolic acidosis with a normal unmeasured plasma anion gap (AG). The major causes of reduced NH_4^+ excretion and hyperkalemia are a low GFR or low aldosterone bioactivity.

5. With hyperkalemia, HCO_3^- can be given safely if indicated (with hypokalemia, HCO_3^- therapy may be dangerous if not accompanied by aggressive K^+ replacement). In metabolic acidosis with a normal plasma AG, HCO_3^- therapy may be more essential than in patients in whom the plasma AG is wide, because metabolism of the unmeasured anion in the latter case leads to HCO_3^- generation.

6. No. Acidemia is NOT always present during metabolic acidosis; a coexistent metabolic or respiratory alkalosis may lower the $[H^+]$ of the plasma to a normal or lower than normal value in a patient with metabolic acidosis.

7. Yes. If the P_{CO_2} is lowered by hyperventilation, the HCO_3 buffer equation is shifted to the right, lowering both the $[H^+]$ and $[HCO_3^-]$. This is called respiratory alkalosis.

8. Yes. Renal HCO_3^- generation is essential to maintain acid-base balance only if there is ongoing net HCO_3^- consumption (i.e., a dietary H^+ load). If one consumes a diet that does not present an acid load (e.g., certain vegetarian diets), acid-base balance is maintained without renal HCO_3^- generation; in fact, maintaining this balance requires HCO_3 excretion in the urine.

DEFINITION

> METABOLIC ACIDOSIS IS PRESENT WHEN PLASMA $[H^+]$ IS ELEVATED AND PLASMA $[HCO_3^-]$ IS DECREASED

In metabolic acidosis, the $[H^+]$ is increased together with a decline in the plasma $[HCO_3^-]$. Two processes can cause metabolic acidosis: a gain of an acid or a loss of $NaHCO_3$ (Table 3–1). Making this distinction is the first step to take in the differential diagnosis of metabolic acidosis.

DEVELOPMENT OF METABOLIC ACIDOSIS

> - THE ACID-BASE STATUS IS DETERMINED BY THE NET BALANCE BETWEEN HCO_3^- PRODUCTION AND LOSS
> - THE GENERAL CAUSES OF METABOLIC ACIDOSIS ARE
>   - HCO_3^- LOSS (ACID OVER-PRODUCTION, $NaHCO_3$ LOSS)
>   - FAILURE OF RENAL *NEW* HCO_3^- GENERATION

We have classified metabolic acidosis in terms of HCO_3^- loss and failure of renal *new* HCO_3^- generation (Table 3–1). The individual diseases are discussed later.

RESPIRATORY RESPONSE DURING METABOLIC ACIDOSIS

> - ACIDEMIA STIMULATES VENTILATION, LOWERING THE P_{CO_2} AND REDUCING THE ECF $[H^+]$ AT A GIVEN $[HCO_3]$
>
> $$H^+ + HCO_3^- \rightleftharpoons H_2CO_3 \rightleftharpoons H_2O + CO_2 \quad (1)$$
>
> - FOR EVERY 1 mM FALL IN $[HCO_3^-]$ *FROM 25 mM* IN METABOLIC ACIDOSIS, THE P_{CO_2} SHOULD FALL 1 mm Hg *FROM 40 mm Hg*

Acidemia is a potent stimulus to the respiratory center. In metabolic acidosis, hyperventilation (lowering the P_{CO_2})

TABLE 3–1. Overview of the Etiology of Metabolic Acidosis

Bicarbonate Loss

- ***Acid Overproduction (Wide Plasma AG)****
  - Endogenous
    - Lactic acidosis (lactic acid)
    - Ketoacidosis (largely β-hydroxybutyric acid)
    - Organic acid overproduction in the GI tract (D-lactic acidosis)
  - Exogenous Intoxications
    - Methyl alcohol (formic acid, lactic acid)
    - Ethylene glycol (glyoxalic acid, lactic acid)
    - Salicylates (lactic and ketoacids plus some acetosalicylate)
    - Paraldehyde (acetic acid)
- ***Actual Bicarbonate Loss (Normal Plasma AG)***
  - Gastrointestinal Tract
    - Diarrhea
    - Ileus
    - Fistula drainage
    - T-tube drainage
    - Villous adenoma
    - Ileal conduit (combined with renal chloride delivery)
  - Urinary Tract
    - Proximal renal tubular acidosis
    - Carbonic anhydrase inhibitors
    - Acid production and urinary organic anion loss

Failure of Renal New Bicarbonate Generation (Low NH_4^+ Excretion)

- Low NH_4^+ and HCO_3 Production
  - Renal failure (low GFR)
  - Hyperkalemia
- Low NH_4 Transfer to the Urine
  - Low distal net H^+ secretion
  - Medullary interstitial disease

*AG = plasma unmeasured anion gap. Each disorder is discussed in more detail in later sections.

displaces the equilibrium of the HCO_3^- buffer system (Equation 1) to the RIGHT and LOWERS the ECF $[H^+]$. Empirically, there is a LINEAR relationship between the plasma $[HCO_3^-]$ and P_{CO_2} in patients with METABOLIC ACIDOSIS. As the ECF $[HCO_3^-]$ falls, the P_{CO_2} also falls, and the slope of the line is approximately 1, which means that the P_{CO_2} falls 1 mm Hg *from 40 mm Hg* for every 1 mM fall in the plasma $[HCO_3^-]$ *from 25 mM*. This linear relationship provides the clinician with a useful tool to assess patients with metabolic acidosis. If the P_{CO_2} is significantly lower than predicted by the above relationship, the patient has another primary stimulus to the respiratory center in addition to the acidemia (respiratory alkalosis). Similarly, if the P_{CO_2} is significantly higher than predicted, there is a compromised ability to ventilate in response to normal stimuli (respiratory acidosis). The importance of the additional disturbance is seen in the following examples:

Patient	A	B	C
$[HCO_3^-]$ mM	8	8	8
P_{CO_2} mm Hg	15	23	34

Questions (Answers on page 47)

9. What is the plasma $[H^+]$ in each case?
10. What is the contribution of the respiratory disorder to this $[H^+]$ in each case?
11. Does the respiratory disturbance influence your approach to the patient's management?

Detrimental Effect of Inadequate Hyperventilation

The importance of adequate hyperventilation during metabolic acidosis is evident in the answers to questions 9–11. Thus, patients with significant chronic lung disease or suppressed ventilation for other reasons are at a major disadvantage if they develop moderately severe metabolic acidosis and are incapable of lowering their P_{CO_2} appropriately. This can occur because of central suppression of ventilation or a mechanical problem (airway obstruction or respiratory muscle failure). The consequences of the failure of hyperventi-

Answers

9. The $[H^+]$ is calculated as follows: ($[H^+] = 23.9 \times Pco_2/[HCO_3^-]$). The respective $[H^+]$ are (A) 47 nM, (B) 72 nM, and (C) 106 nM (the respective pH values are 7.33, 7.14, and 6.97).

10. The fall in $[HCO_3]$ from 25 mM should equal the fall in Pco_2 from 40 mm Hg (40 − 17 = 23). Therefore, Case B represents the appropriate respiratory response to metabolic acidosis; Case A represents combined metabolic acidosis and respiratory alkalosis, whereas Case C demonstrates metabolic acidosis and respiratory acidosis. Therefore, with superimposed respiratory alkalosis, the $[H^+]$ is reduced by 25 nM (72 − 47 nM), whereas with superimposed respiratory acidosis, the $[H^+]$ is increased by 34 nM (106 − 72 nM).

11. The superimposed respiratory acidosis has turned Case C into a life-threatening acidemia, and intervention is indicated. Therapy should be aimed at the reversible components of both the metabolic and the respiratory acidoses. The patient requires mechanical ventilation when the Pco_2 is high enough to cause very severe acidemia (this changes the $[H^+]$ rapidly).

TABLE 3–2. Plasma $[H^+]$ in Patients With Progressive Metabolic Acidosis With and Without Respiratory Compensation*

Plasma $[HCO_3$ (mM)$]$	Status With Appropriate Hyperventilation		Status Without Hyperventilation	
	$[H^+]$ (nM)	*pH* (Units)	*$[H^+]$* (nM)	*pH* (Units)
20	42	7.38	48	7.32
15	48	7.32	64	7.19
10	60	7.22	96	7.02
5	96	7.02	191	6.72

*Patients with respiratory compensation have a 1 mm Hg reduction in Pco_2 for each mM fall in HCO_3 concentration. Patients without respiratory compensation have a Pco_2 of 40 mm Hg in each case.

lation are depicted in Table 3–2, page 47. The impact of hyperventilation on the defense of $[H^+]$ must be kept in mind whenever the patient with metabolic acidosis is subjected to any procedure that may interfere with ventilation, e.g., sedation to carry out an invasive procedure such as gastroscopy. Furthermore, when a patient with metabolic acidosis is given an anesthetic, ventilation must be continued at the level appropriate for that patient's degree of metabolic acidosis, otherwise severe acidemia will result.

DIAGNOSTIC TESTS TO RESOLVE THE BASIS OF THE METABOLIC ACIDOSIS

Plasma Unmeasured Anion Gap (AG)

$$\text{PLASMA AG} = [Na] - ([Cl] + [HCO_3]) = 12 \pm 2 \text{ mEq/L}$$

UNMEASURED ANIONS ARE USEFUL TO
- IDENTIFY, CLASSIFY, AND MONITOR METABOLIC ACIDOSIS
- QUANTITATE ORGANIC ANIONS; 1 mM FALL IN PLASMA $[HCO_3]$ LEADS TO A 1 mEq/L RISE IN PLASMA AG

The number of positive and negative charges are equal in the ICF and the ECF. Nevertheless, when one calculates the difference between the major univalent plasma cation Na and anions Cl and HCO_3, the value is POSITIVE (12 ± 2 mM) and is referred to as the UNMEASURED ANION GAP (AG). This is a *calculation of diagnostic convenience* that has several applications during metabolic acidosis (see page 49 for details):

1. Identification of patients with metabolic acidosis.
2. Diagnostic classification of metabolic acidosis based upon acid gain (wide AG) vs $NaHCO_3$ loss (normal AG) (Table 3–3, page 49).
3. Convenient monitoring of the course of this metabolic disorder.
4. Quantitative assessment of the magnitude of the added organic acid load.
5. Detection of laboratory errors.

TABLE 3–3. Metabolic Acidosis Classified According to the Plasma AG

Increased AG	*No Increase in AG*
Lactic Acidosis	$NaHCO_3$ Loss
Ketoacidosis	• Direct (GI or Renal)
Renal Failure	• Indirect
Intoxications	Low urine [NH_4^+]
Organic Acids From the GI Tract	Organic anion excretion
(e.g., D-Lactic Acid)	$NaHCO_3$ Dilution
	Organic Acidosis With Narrow Normal AG
	(e.g., Hypoalbuminemia)

Explanation of the Term Plasma "Unmeasured Anion Gap"

We use the term plasma unmeasured anion gap (AG) to account for the difference between the [Na^+] and the sum of [Cl^-] + [HCO_3^-]. This shortfall in the number of anions is due to the fact that a significant number of the anions are unmeasured—hence, the term "unmeasured anion gap." In truth, there are a large number of both unmeasured cations and anions, and the plasma AG reflects the difference between the unmeasured cations (K, Mg, and Ca) and the unmeasured anions (albumin, organic anions, PO_4, and SO_4).

Genesis of the Unmeasured Anions in Metabolic Acidosis

When lactic acid dissociates into H^+ and lactate anions in the ECF, H^+ are buffered by HCO_3^-, leaving lactate anions as the "footprint" of the lactic acid added to the ECF. These events are depicted below with values before and after the addition of 10 mmol of lactic acid (HL) to each L of the ECF.

Plasma	Na	Cl	HCO_3	AG
Normal	140	103	25	12
+10 mM HL	140	103	15	12+10

The patient begins with a normal plasma AG of 12 mEq/L (140 − [103 + 25]). The protons from 10 mM lactic acid react with ECF HCO_3^-, resulting in a fall of plasma [HCO_3] to 15 mM and an increase in the AG to 22 mM, which is best thought of as the increment over normal, i.e., 12 + 10 (this forces you to think of the normal plasma AG and to ask whether the patient has any reason not to have a normal AG; see page 51). Thus, the increase in the AG reflects, in a quantitative fashion, the presence of the anion of the organic acid (in this case, 10 mM lactate).

If, in the above example, the increase in the plasma AG were the same 10 mEq/L but the plasma [lactate] were only 5 mM, lactic acidosis would not be the sole cause of the metabolic acidosis; the patient must also have accumulated another unmeasured anion (e.g., ketoacid anions, Table 3–1).

The AG of 12 mEq/L indicates the presence of unmeasured anions in the plasma. Some other ions are present in small concentrations and vary little in most circumstances; thus, they need not be considered in the calculation of the plasma AG (K, Ca, Mg, PO_4 and SO_4). The anion that is "unmeasured" under normal circumstances is primarily albumin.

Sometimes, the accumulation of unmeasured anions is NOT apparent because the normal AG is abnormally narrow (see page 51 for details).

In patients with a metabolic acidosis associated with an increase in AG, there are 2 possible bases for the metabolic acidosis:

1. Overproduction of an organic acid.
2. Renal failure (low GFR).

Overproduction of an Organic Acid. Patients may produce organic acids as a result of the excess activity of a normal metabolic pathway. For example, there is an exceedingly rapid rate of lactic acid production during hypoxia and a quite rapid rate of ketoacid production during states of relative insulin deficiency (see page 9 for more details). In addition, the metabolism of an ingested substance (methyl alcohol, ethylene glycol, paraldehyde) may lead to a rapid rate of acid production (Table 3–1, page 45). In salicylate intoxication, a small increase in unmeasured anions may develop, owing to dissociation of the salicylic acid, and if metabolic acidosis is present, there will be a much larger increase in the plasma AG, owing to the production of other organic acids.

Renal Failure. The increase in plasma AG in renal failure is illustrated in Figure 3–6, page 77.

Metabolic Acidosis and a Normal Plasma AG

When metabolic acidosis is *not* associated with an increase in the plasma AG, it is due to $NaHCO_3$ loss, which may be direct or indirect (Table 3–3, page 49). Direct HCO_3^- loss occurs either via the gastrointestinal tract or in the urine; indirect HCO_3^- loss occurs because of the failure of the kidneys to excrete NH_4^+ and to generate *new* HCO_3.

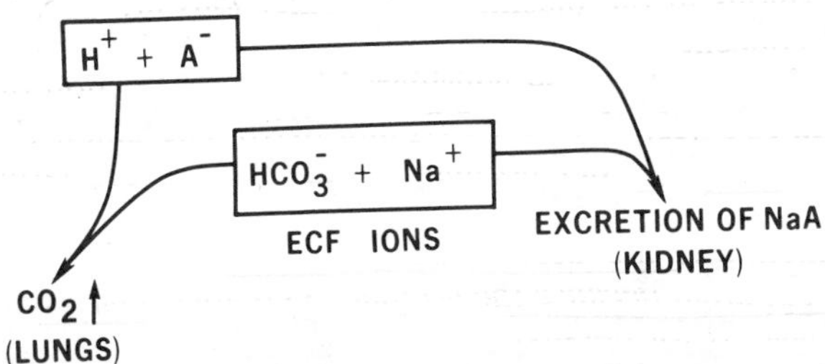

Figure 3–1. Metabolic acidosis with a normal plasma AG in ketoacidosis. The larger rectangle represents the ECF and contains $NaHCO_3$. β-hydroxybutyric acid is produced in the liver (smaller rectangle). The H^+ so produced combine with ECF HCO_3^-, yielding CO_2, which is eliminated via the lungs (loss of ECF HCO_3^-). The β-hydroxybutyrate anion is filtered by the kidney and excreted in the urine. If this filtered load is so high that it exceeds the normal kidney's ability to excrete NH_4^+ (or if a reduced rate of NH_4^+ excretion is present [distal RTA]), the net result is the excretion of Na^+ (+ K^+) β-hydroxybutyrate in the urine (loss of ECF Na). The above two processes lead to the net loss of $NaHCO_3$ from the ECF.

CAUSES OF A REDUCED PLASMA AG

Hypoalbuminemia. Albumin appears to have a net anionic equivalence of close to 12 mEq/L. Therefore, if a patient has a serum albumin of 2 gm/dl or 20 gm/L (half the normal concentration), the AG is reduced by close to 6 mEq/L.

Increased Unmeasured Cations. Patients with lysine- or arginine-rich proteins in their plasma (IgA myeloma) have an abnormal paraprotein that carries a net positive charge, and if high enough, it can actually render the plasma AG NEGATIVE.

Hypercalcemia generally is not severe enough to have a significant impact on the plasma AG.

Laboratory Errors. Patients may have an unexpectedly low plasma AG due to quirks of the laboratory methods or actual laboratory errors. In patients with HALIDE INTOXICATION (bromide or iodide) and HYPERLIPIDEMIA (with techniques depending on turbidity), the plasma [Cl] will be overestimated. Similarly a simple error in the measurement of [Na], [Cl] or [HCO_3^-] will make the calculation of the plasma AG invalid.

Excretion of the Organic Anion. Metabolic acidosis of the increased AG type might not be associated with an actual increase in the plasma AG if the anion is excreted in the urine at an abnormally high rate. This has been observed particularly in diabetic ketoacidosis, in which the ketonuria is so great that there is a substantial indirect loss of $NaHCO_3$ (Fig. 3–1 above).

Tests to Assess Renal Contribution to Acid-Base Balance

- **URINE pH:** GOOD FOR BICARBONATURIA
  POOR FOR NH_4^+ EXCRETION
- **URINE NET CHARGE:** GOOD FOR NH_4^+ EXCRETION
- **ALKALINE URINE P_{CO_2}:** GOOD FOR COLLECTING DUCT H^+ SECRETION

Renal HCO_3^- generation depends on 2 processes: collecting duct H^+ secretion and glutamine metabolism plus NH_4^+ excretion. One can assess these processes indirectly, using the following tests.

Urine pH. The urine pH has 1 major application in the assessment of acid-base disorders: it provides an excellent means to detect the presence of bicarbonaturia (pH >7 indicates $[HCO_3] > 10$ mM).

The urine pH was also used to assess *new* HCO_3 formation by the kidney (i.e., NH_4^+ excretion), but it is inadequate for that purpose. As indicated in Figure 3–2, in acute acidemia (left panel), NH_4^+ excretion increases as the urine pH falls (more NH_4^+ trapped in the lumen), whereas in chronic acidemia, the converse is true, owing to the markedly enhanced NH_3 availability. As evident from Figure 3–3, a defect in NH_3 availability results in low NH_4^+ excretion and a *low urine pH*, whereas a defect in H^+ secretion results in a low NH_4^+ excretion and a *high urine pH*. Thus, without a simultaneous assessment of the urine $[NH_4^+]$, the urine pH is a potentially misleading parameter to reflect *new* renal HCO_3^- formation.

Urine P_{CO_2} (in alkaline urine). If the P_{CO_2} of alkaline urine is >70 mm Hg, collecting duct (CD) H^+ secretion is normal if the test has been done correctly (see page 55 for details). If the urine P_{CO_2} is <50 mm Hg, there is impaired H^+ secretion. The urine P_{CO_2} is measured in alkaline urine (i.e., luminal $[H^+]$ lower than ICF $[H^+]$). Therefore, in conditions in which net H^+ secretion is impaired because there is backleak of H^+ from lumen to cell (e.g., amphotericin B), the urine P_{CO_2} should be normal in the presence of other tests, indicating low NH_4^+ excretion (see page 55 for rationale).

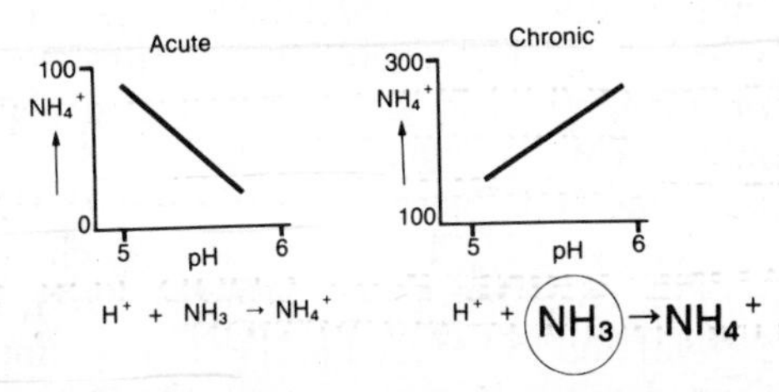

Figure 3–2. Urine pH and NH_4 excretion. In acute acidosis, the rate of NH_4 excretion is higher when the urine pH is *lower*, the increased $[H^+]$ trapping more NH_3 in the lumen as NH_4^+. Thus the H^+ secretion rate exceeds that of NH_3.

In chronic metabolic acidosis, both the H^+ secretory rate and the NH_3 availability are increased. The increase in NH_3 is due to augmented ammoniagenesis and is relatively larger than the increment in H^- secretion. Thus, the urine NH_4^+ excretion rate is increased in conjunction with a higher urine pH.

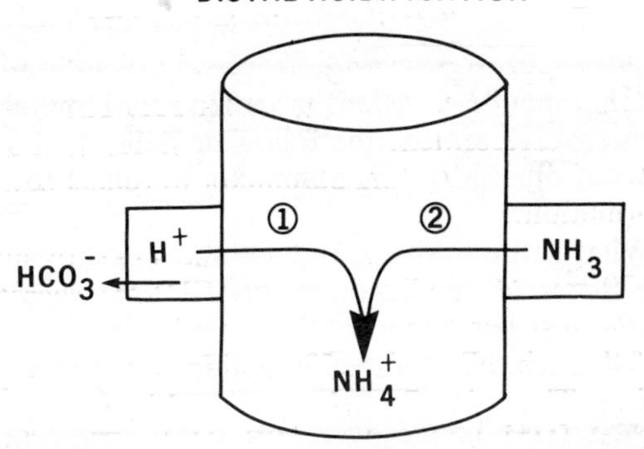

Figure 3–3. Model of H^+- and NH_3-secretion in the collecting duct. A schematic to illustrate that the urine pH can be high with a low NH_4^+ excretion rate if the decrease is primarily due to a reduced H^+ secretion (1). In contrast, the urine pH can be low with a low NH_4^+ excretion rate if the decrease is primarily due to a reduced NH_3 availability (2).

Tests to Assess NH_4^+ Excretion. Although the urine $[NH_4^+]$ can be assayed directly, this test is generally not available in the routine clinical laboratories. However, a knowledge of urine NH_4^+ excretion is critical to diagnose the basis of metabolic acidosis with a normal plasma AG. This information can be gained as follows:

Apparent Urine Net Charge. NH_4^+ is a CATION and is usually excreted as NH_4Cl. Under normal circumstances with a urine pH <6.1 (hardly any urine HCO_3^-), the urine CATIONS ([Na] and [K] exceed [Cl] (owing to unmeasured anions, PO_4 and SO_4). However, when the $[NH_4^+]$ is high, the urine [Cl] exceeds the [Na] + [K], indicating an unmeasured cation (i.e., NH_4^+). In quantitative terms, the urine $[NH_4^+]$ = 80 mM when the urine [Cl] = ([Na] + [K]), providing the urine volume is not excessive.

The urine net charge does not reveal a high $[NH_4^+]$ if the NH_4^+ is excreted in conjunction with an UNMEASURED ANION (e.g., β-hydroxybutyrate). The presence of this unmeasured complex is revealed by comparing the measured urine osmolality with that calculated from the urine Na + K + Cl + urea and glucose + 90, all in mM terms. This is called the urine osmolal gap (see page 97 for details).

THERAPY FOR METABOLIC ACIDOSIS—GENERAL CONSIDERATIONS

> The therapeutic decisions in patients with metabolic acidosis revolve around the following issues:
>
> 1. What emergency measures are required for survival?
> 2. What is the basis for the metabolic acidosis?
> 3. Is there an important reversible underlying disorder?
> 4. What are the therapeutic options for treating the acidosis per se?

Emergency Measures. Before the biochemical results are available, measures to ensure a proper airway, adequate circulation, and oxygen delivery must be pursued with vigor.

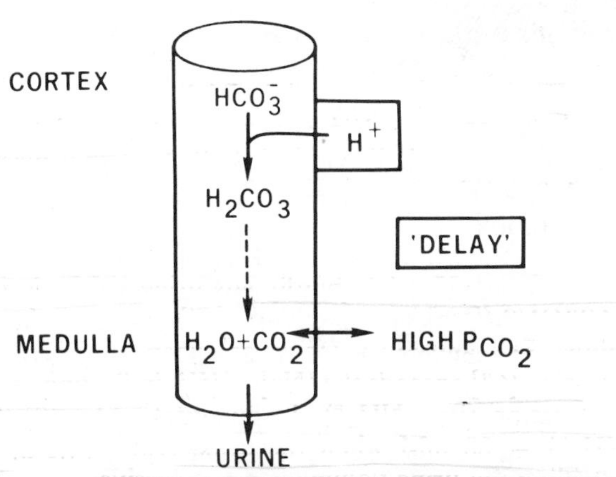

Figure 3–4. The basis of increased alkaline urine P_{CO_2}. Because the urine is alkaline, any H^+ secretion in the collecting duct (CD) leads to the formation of H_2CO_3. As there is no carbonic anhydrase present in this nephron segment, the H_2CO_3 dehydrates slowly in the medullary collecting duct, leading to an increase in the urine P_{CO_2}. If CD H^+ secretion is absent, the urine P_{CO_2} will be close to that of the blood.

URINE P_{CO_2}, A TEST OF COLLECTING DUCT H^+ SECRETION

This test requires the administration of sufficient $NaHCO_3$ to render the urine frankly alkaline and therefore also allows you to rule out pRTA (the blood [HCO_3^-] becomes normal before marked bicarbonaturia develops). The test is valid only once the urine pH is >7.40 or the urine [HCO_3^-] is >50 mM. This can usually be achieved by giving the patient an oral $NaHCO_3$ load of 0.5 to 2 mmol/kg body weight on the morning of the test. The patient's K^+ deficit should be corrected prior to giving HCO_3^-.

The urine is collected in a bottle with a small surface area relative to volume to minimize CO_2 loss. No oil is necessary. The sample is aspirated from the bottom of the bottle into a sealed syringe. The pH and P_{CO_2} are measured anaerobically.

The patient with normal collecting duct H^+ secretion should have a urine P_{CO_2} exceeding 70 mm Hg. This test also enables the identification of patients with a gradient limit to H^+ secretion (back diffusion of H^+, e.g., owing to use of amphotericin B): these patients have low urine NH_4^+ excretion but have a normal urine P_{CO_2}.

In a very alkaline urine, H^+ secreted by the collecting duct lead to the formation of H_2CO_3. As there is no luminal carbonic anhydrase here, H_2CO_3 is slowly dehydrated to CO_2 in the medulla, leading to an increased P_{CO_2} of bladder urine. Patients with a H^+ secretory defect have a urine P_{CO_2} close to that of their blood (Fig. 3–4).

Specific Basis for Metabolic Acidosis. There are 3 critical reasons to make a specific diagnosis:

1. To determine whether the rate of H^+ production is so high that arresting H^+ production is the major effective measure (e.g., diabetic ketoacidosis). In contrast, if the rate of H^+ production is likely to be low (renal causes), other considerations become preeminent.

2. The cause of the metabolic acidosis may pose a serious but independent threat to the patient (e.g., methanol overdose). Its specific therapy (ethanol administration) is the most important therapeutic measure.

3. Certain types of metabolic acidosis are associated with hypokalemia (low distal H^+ secretion, diarrhea). In this case, K replacement may have to be initiated before HCO_3 administration in order to avoid serious cardiac arrhythmias or respiratory failure.

Underlying Disorders. For the sake of brevity, these are not discussed here. However, their therapy may determine the eventual outcome (e.g., sepsis).

Therapeutic Options. The above having been treated, the specific therapeutic options are as follows:

1. *Stop H^+ production:* This is critical in conditions with a very rapid rate of H^+ production. H^+ production can be 60 mmol/min in lactic acidosis (page 9) or 10 mmol/min in diabetic ketoacidosis; it can also be very rapid with methanol overdose. Therefore, specific therapy such as oxygen in lactic acidosis due to hypoxia, insulin in diabetic ketoacidosis, ethanol in methyl alcohol intoxication, and possibly gastric lavage in certain intoxications can be life-saving.

2. *Ventilation:* To lower the plasma $[H^+]$ in a period of minutes, the only nonspecific therapeutic option is to ensure an adequate degree of hyperventilation. Therefore, this option is most useful in combined respiratory and metabolic acidosis and is the initial treatment of choice in this case (see Table 3–4, page 59).

3. *Increase endogenous HCO_3^- formation:* This option for therapy is only available in patients with a wide AG type of metabolic acidosis. The only emergency measure of value is to increase the metabolism of the circulating organic anions (renal new HCO_3 formation can only occur at a rate of 0.3 mmol/min with perfectly adapted kidneys). Net removal of lactate and β-hydroxybutyrate requires you to stop their production (see Paragraph 1 above). In lactic acidosis

BASICS OF HCO_3 THERAPY

When calculating the amount of $NaHCO_3$ to administer, assume a volume of distribution of 50% of total body weight (TBW). This figure is derived from experimental data and reflects the fact that 60% of buffering occurred in the ICF. Initial HCO_3^- therapy should aim to remove the patient from immediate danger, i.e., to raise the plasma [HCO_3^-] to 10 mM. The rapidity with which the $NaHCO_3$ is given is determined by the severity of the acidemia as well as the K status and the cardiac status of the patient. The tonicity of the $NaHCO_3$ administered should be determined by the patient's tonicity.

With very low plasma [HCO_3^-], the volume of distribution of HCO_3^- may exceed 50% of TBW, owing to increased ICF H^+ buffering.

The administration of $NaHCO_3$ in an acidemic patient is a CO_2-producing process, and if the patient is being ventilated, the P_{CO_2} rises if the alveolar ventilation is not increased.

THERAPEUTIC OPTIONS IN PATIENTS WITH METABOLIC ACIDOSIS, RENAL FAILURE, AND ECF VOLUME EXPANSION

Gastric HCO_3^- Generation. The stomach usually generates 150 mmol of HCO_3^- daily, and this can be mobilized as a therapeutic tool in these challenging patients. One can insert a nasogastric tube, stimulate gastric acid secretion with pentagastrin (provided that the patient has not received histamine receptor blockers and is not achlorhydric), and remove significant amounts of HCl. One must ensure that the nasogastric tube is well situated to remove most of the acid. The periodic instillation of antacids down the tube helps to prevent complications secondary to excess acid secretion. Should sufficient Na be removed via this route, some $NaHCO_3$ may be administered intravenously as well.

Phlebotomy and Dialysis. If the patient with severe metabolic acidosis is in frank pulmonary edema, in addition to the usual therapeutic maneuvers (diuretics, oxygen, morphine, digoxin), one should do a phlebotomy early to allow the administration of $NaHCO_3$, as acidemia may also impair cardiac function. The phlebotomized blood should be packed and the cells returned to the patient. If the patient has renal failure, early dialysis with a HCO_3^- bath should be planned.

Ventilation. One should consider ventilating these patients early to lower their P_{CO_2} appropriately to treat their acidemia. One can influence the acid-base state much faster by lowering the P_{CO_2} than by the administration of HCO_3^-. In patients with pulmonary edema, ventilation is also beneficial for the pulmonary edema if the patient can tolerate positive end-expiratory pressure.

that is not due to anoxia, the activator of pyruvate dehydrogenase, dichloroacetate, may be of temporary value.

4. *Exogenous $NaHCO_3$ administration:* The rationale for administering large quantities of HCO_3 is to treat serious acidemia (see Tables 3–4 through 3–6, page 59) or to "buy a little time" when the rate of H^+ production can only be arrested with a little delay (e.g., insulin in very severe diabetic ketoacidosis). The danger of this therapy is primarily the Na load.

Recall that to raise the plasma [HCO_3] by 5 mM, you have to give 150 mmol of HCO_3 plus enough HCO_3 to titrate the ongoing rate of H^+ production. The patient with renal failure and severe acidemia is especially threatened by this problem. If the [HCO_3^-] is <8 mM you should try to raise the plasma [HCO_3] with $NaHCO_3$ to 10–12 mM.

We recommend that the plasma [HCO_3] rather than the pH be used to assess the need for $NaHCO_3$ therapy in order to avoid being misled by acute hyperventilation (Table 3–6). One should liberalize these criteria in the patient with chronic lung disease whose ability to hyperventilate during metabolic acidosis might be severely compromised (Table 3–4).

Another important consideration is hypokalemia. Exogenous $NaHCO_3$ can lead to a K shift into cells and thereby induce respiratory failure or an arrhythmia. The factors putting patients at risk of coexistent K depletion and metabolic acidosis are listed in Table 3–7A. The therapeutic approach to these patients is detailed in Table 3–7B.

Bicarbonate Therapy in Metabolic Acidosis

1. HCO_3^- INDICATED IN THE FOLLOWING:
   - PLASMA [HCO_3] <5 mM
   - PROBLEM WITH ADEQUATE HYPERVENTILATION
   - PATIENTS WITH SEVERE METABOLIC ACIDOSIS AND NORMAL AG
   - SEVERE ACIDEMIA AND RENAL FAILURE OR INTOXICATION
2. HCO_3^- DISTRIBUTES IN VOLUME EQUAL TO 50% OF BODY WEIGHT
   - LARGER WHEN PLASMA [HCO_3^-] <5 mM

TABLE 3–4. Acid-Base Status of the Patient With Chronic Pulmonary Disease and Superimposed Metabolic Acidosis*

State	$[H^+]$ (nM)	pH	P_{CO_2} (mm Hg)	$[HCO_3^-]$ (mM)
Chronic respiratory acidosis	47	7.31	60	31
Superimposed metabolic acidosis	150	6.80	60	10
Metabolic acidosis with the expected respiratory response	63	7.20	25	10

*In chronic lung disease, normal ventilatory compensation for metabolic acidosis cannot be assumed.

TABLE 3–5. Impact of Small Changes in $[HCO_3^-]$ or P_{CO_2} on the Acid-Base Status of the Patient With a Plasma $[HCO_3]$ of 7 mM

	$[H^+]$ (nM)	pH	P_{CO_2} (mm Hg)	$[HCO_3^-]$ (mM)
A. Stable metabolic acidosis	75	7.12	22	7
B. Small reduction in hyperventilation	120	6.99	30*	7
C. Small fall in plasma $[HCO_3^-]$	96	7.02	20	5*

*Note that the slight increase in P_{CO_2} in B or the small fall in $[HCO_3^-]$ in C converts a modest acidemia into a severe one.

TABLE 3–6. "Acceptable" pH Due to Acute Hyperventilation, Which Cannot Be Maintained Chronically

	$[H^+]$ (nM)	pH	P_{CO_2} (mm Hg)	$[HCO_3^-]$ (mM)
ACUTE HYPERVENTILATION	48	7.32	10	5
CHRONIC STATE	96	7.02	20	5

TABLE 3–7A. Potassium Depletion and Metabolic Acidosis

Cause	Basis of Hypokalemia
RTA (low H^+ secretion type)	Renal K wasting
GI $NaHCO_3$ loss	GI K loss, renal K loss
Diabetic ketoacidosis*	Osmotic diuresis

*The K-depleted diabetic may actually be hyperkalemic at presentation, owing to a shift of K from the ICF to the ECF with insulin deficiency.

TABLE 3–7B. Principles of Therapy in Acidemic K^+-Depleted Patients

Use Oral Route for K^+ Whenever Possible	Can give large amounts of K^+, avoiding the IV problems.
Use several IV Sites	Allows the dissociation of K and HCO_3^- infusion rates, and permits more aggressive K administration by peripheral vein.
Use Cardiac Monitor	Early detection of arrhythmias.
Avoid Central K^+ Bolus Whenever Possible	Renders intracardiac blood very hyperkalemic and may produce serious arrhythmias.
Ensure Adequate Urine and K^+ Output (>30 ml/hr, >20 mM $[K^+]$)	If the patient has renal failure (rule out prerenal), give K^+ more cautiously.

DIAGNOSTIC APPROACH TO METABOLIC ACIDOSIS

The diagnostic approach to the patient with metabolic acidosis is outlined in the flow charts on pages 61, 62, and 63. Those patients with an increase in the plasma AG are classified in Table 3–3, page 49, and the chart on page 61 is self-explanatory.

In patients with metabolic acidosis and no increase in the plasma AG (flow chart 3–B, page 62), begin by ruling out an organic acid acidosis that presents with no "apparent" increase in the plasma AG because of laboratory "error."

One uses the urine net charge to reflect urine [NH_4^+]. A nonrenal basis for metabolic acidosis (e.g., diarrhea) is suggested when the urine NH_4^+ excretion is >100 mmol/day. In this case, the urine [Cl] will exceed the urine [Na] + [K]. In contrast, with a low urine [NH_4^+], suspect RTA (urine [Na] + [K] exceed [Cl]). Rarely, NH_4^+ may be in the urine in conjunction with an anion other than Cl (β-hydroxybutyrate), the latter representing a subgroup of patients with ketoacidosis and no increase in the plasma AG due to the marked ketonuria. This excretion may be revealed by calculating the urine osmolal gap (page 97).

The diagnostic approach to the patient with RTA (identified by the low urine [NH_4^+], as on page 54) is summarized in the flow chart 3–C on page 63. One begins by separating proximal from distal RTA by detecting bicarbonaturia during metabolic acidosis, using the urine pH. The distal H^+ secretory mechanism is evaluated by 2 tests. If the urine P_{CO_2} is 30 mm Hg above that of the blood, the distal H^+ secretory mechanism is normal; a urine P_{CO_2} of 55 mm Hg or less indicates a H^+ pump defect (see page 55 for details). In patients with a normal urine P_{CO_2}, a urine pH <5 suggests a defect in NH_4^+ production or NH_3 transfer in the medulla, whereas values >6 suggest a defect in collecting duct H^+ secretion (backleak in view of normal urine P_{CO_2}). Hyperkalemia indicates those patients with low aldosterone bioactivity.

Patients can have dRTA with normal H^+ secretion for 1 of 2 reasons: either there is a defect in NH_3 availability or else the collecting duct is abnormally permeable to H^+, allowing H^+ backleak from the lumen.

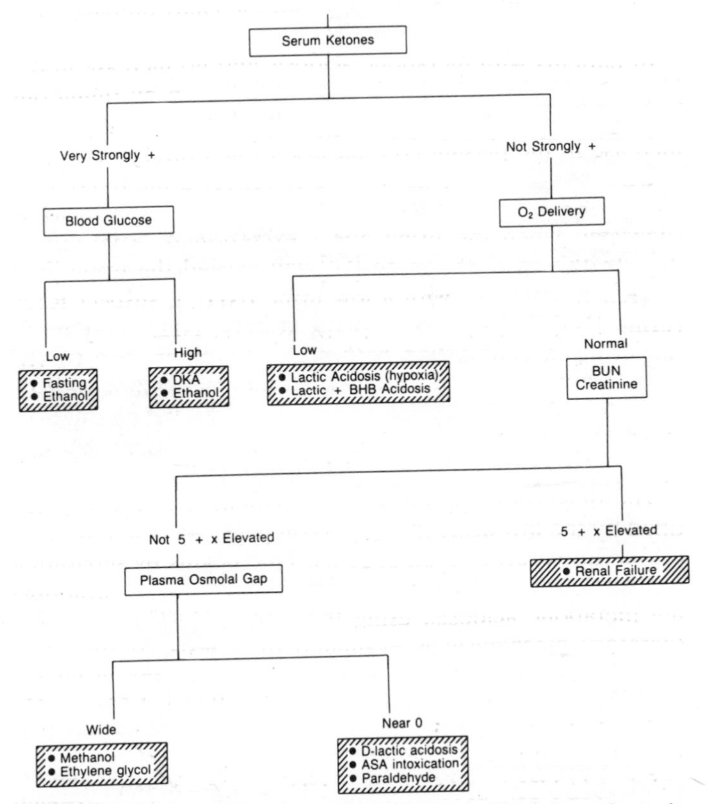

Flow Chart 3–A. Approach to metabolic acidosis. The information to be interpreted to make a final diagnosis is shown in the open boxes; the final diagnoses are show in the shaded boxes. The first critical question is "Is the plasma anion gap wide?"

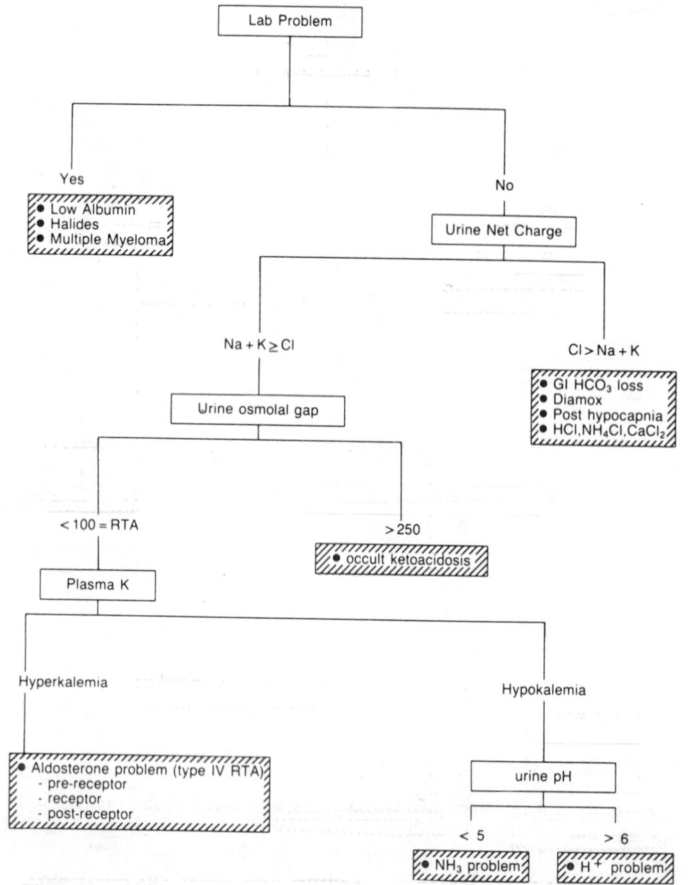

Flow Chart 3–B. Approach to the patient with metabolic acidosis and a normal plasma AG. The information to be interpreted to make a final diagnosis is shown in the open boxes; the final diagnoses are shown in the shaded boxes.

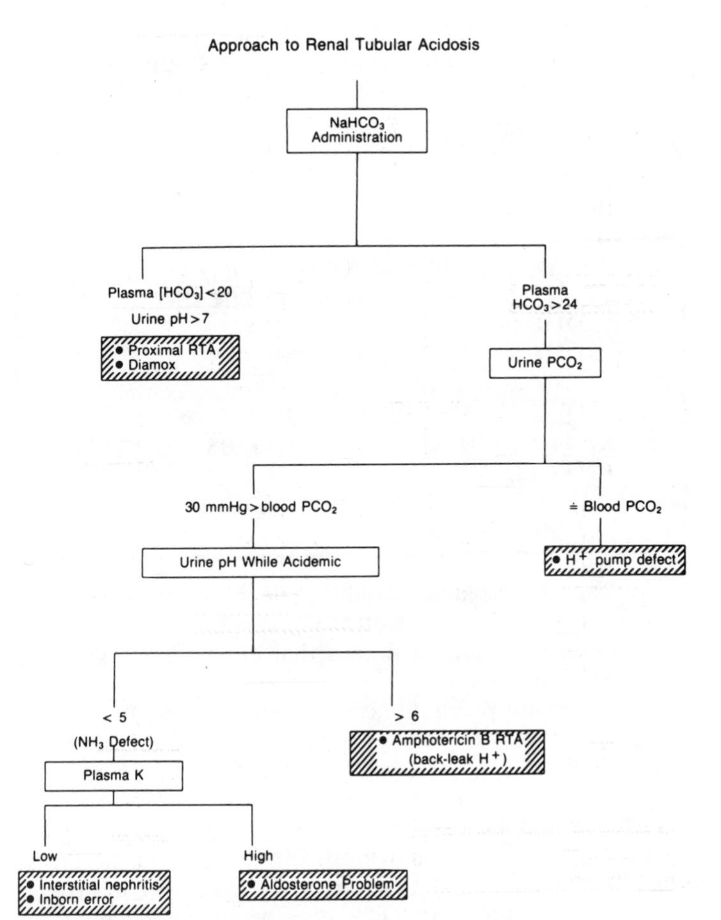

Flow Chart 3–C. Approach to patients with RTA. The information to be interpreted to make a final diagnosis is shown in the open boxes; the final diagnoses are shown in the shaded boxes.

SPECIFIC DISORDERS CAUSING METABOLIC ACIDOSIS

Metabolic Acidosis With a Wide Plasma AG

KETOACIDOSIS

The basic biochemical cause for ketoacidosis is relative insulin deficiency. This leads to excessive fat mobilization, oxidation in the liver, and production of ketoacids (the principal one, β-hydroxybutyrate, is actually a hydroxy-acid). The rate of ketoacid production can be very rapid (100 mmol per hr).

A. ***KETOACIDOSIS WITH NORMAL β-CELL FUNCTION* (i.e., LOW INSULIN RELEASE):**
   DUE TO HYPOGLYCEMIA
   DUE TO INHIBITION OF β CELLS
      (α ADRENERGICS)
   DUE TO EXCESSIVE LIPOLYSIS
B. ***KETOACIDOSIS WITH ABNORMAL β-CELL FUNCTION:***
   DUE TO INSULIN-DEPENDENT DIABETES
      MELLITUS

The causes of relative insulin deficiency are listed in Table 3–8. Two groups are evident:

(1) those with normal β cells that either lack a stimulus or are inhibited and

(2) those with β-cell disease (diabetes mellitus).

The patients with normal β cells are outlined briefly in Table 3–8, page 65. Diabetic ketoacidosis is discussed in detail.

Diabetic Ketoacidosis

Although diabetic ketoacidosis often occurs in the previously diagnosed insulin-dependent diabetic (IDDM) (often with a precipitating event), it may be the initial mode of presentation in young IDDM patients. These patients are usually severely hyperglycemic, complaining of polyuria and polydipsia (this aspect of the presentation is discussed in detail in Chapter 12 and is not emphasized in this section). The degree of hyperglycemia depends on glucose intake and, more importantly, on the degree of reduction in GFR (ECF volume contraction). If the plasma $[HCO_3^-]$ is <10 mM, the patient will have severe hyperventilation (deep,

TABLE 3–8. Etiologic Classification of Ketoacidosis

Cause	Special Features	Treatment
I. ***Insulin Deficiency With Normal B Cells***		
HYPOGLYCEMIA		
Fasting	• Never find plasma: [HCO_3] <18 mM or AG >19 mEq/L • Plasma [glucose] not <3 mM	• Glucose cures it quickly
Liver problem, e.g., glycogen storage disease, Type I or a defect in gluconeogenesis (GNG)	• Hypoglycemia can be marked • Plasma [HCO_3] can be <18 mM • If defect in GNG, patient may also have lactic acidosis	• Glucose eliminates ketoacidosis • Special therapy for underlying disease
INHIBITION OF INSULIN RELEASE		
High α adrenergics Ethanol with vomiting, yielding marked ECF volume contraction	• Pernicious vomiting, patient almost in shock • Ketoacidosis may be severe • Mixed acid-base disorders • Blood [glucose] may be low, normal, or high	• NaCl to restore ECF volume • KCl to replace K deficit • B vitamins • Phosphate • Glucose only if hypoglycemic
II. ***B-Cell Destruction (Diabetes Mellitus)***	• Severe hyperglycemia • Severe ketoacidosis • ECF volume very low • K depletion with hyperkalemia • Look for underlying disease and treat it	• Insulin • NaCl, KCl • Threat of hypokalemia in 2 hr • Glucose needed in 6 hr
III. ***Other Causes***		
EXCESSIVE LIPOLYSIS		
After exercise	• High FFA mobilization continues, but oxidation rate slows	• No real danger, no special treatment
Salicylate overdose	• Activates tissue lipase • Hyperventilation • CNS toxicity • K depletion	• Remove ASA by promoting excretion + GI lavage; dialysis may be necessary

regular Kussmaul respirations). As the blood sugar climbs and the acidosis progresses, there are associated disturbances in the mental state, leading to confusion or coma (perhaps also related to the hyperosmolar state). The ketoacidosis usually develops over 12–24 hours of insulin deficiency and may become progressively more fulminant. Less commonly, it may be more slowly progressive. In this latter case, the degree of ECF volume depletion is more severe. Owing to the greater Na loss via the more prolonged osmotic diuresis, these patients are usually severely hypotensive.

The plasma [K] is usually somewhat elevated (close to 5.0 mM). Hyperkalemia is due to insulin deficiency and is aggravated by tissue catabolism and renal failure. Some patients may be hypokalemic if they had prior K loss or a prolonged osmotic diuresis with aldosterone bioactivity.

The degree of acidemia also varies, depending on the degree of the excessive production of ketoacids.

These patients also have hyponatremia for 4 major reasons:

1. Water movement from the ICF of those cells requiring insulin for glucose entry (Chapter 12)
2. Na loss in the urine in the osmotic diuresis
3. Water ingestion in the presence of ADH
4. A laboratory error secondary to hyperlipemia

Although these patients are hyponatremic, they are still hyperosmolar, owing to the hyperglycemia, and they are obviously Na-depleted.

Diagnosis. The diagnosis of diabetic ketoacidosis is usually not a difficult one to establish—it should be ruled out in all patients with metabolic acidosis and an increase in the plasma AG. Hyperglycemia, ketonemia (positive qualitative test for acetoacetate in a serum dilution of 1:8), glycosuria, and ketonuria are sufficient criteria. The fall in plasma [HCO_3^-] should initially approximate the increase in plasma AG (see pitfalls in diagnosis [page 67] and Table 3–9, tracing the change in the plasma AG and plasma [HCO_3] [page 69]).

Treatment. Before dealing with the details of treatment,

Therapeutic Approach
1. Stop H^+ production
2. Reexpand the ECF volume
3. Correct K deficit

NEGATIVE QUALITATIVE TEST FOR KETONES IN PATIENTS WITH KETOACIDOSIS AND COEXISTENT LACTIC ACIDOSIS

In ketoacidosis, the two ketoacids β-hydroxybutyrate and acetoacetate are in equilibrium:

$$\text{Acetoacetate} + \text{NADH} \rightleftharpoons \beta\text{-hydroxybutyrate} + \text{NAD}$$

The nitroprusside qualitative side room test for ketones detects only ACETOACETATE and ACETONE. Therefore, if one has NADH accumulation in mitochondria such as in LACTIC ACIDOSIS or in alcohol metabolism (Fig. 3–5, page 71), it displaces the equilibrium of this equation to the right in favor of β-hydroxybutyrate. Since the nitroprusside test does not detect β-hydroxybutyrate, the results may well be only weakly positive and one may underestimate the ketoacidosis. However, the enzymatic assay for β-hydroxybutyrate is reliable.

PITFALLS IN THE DIAGNOSIS OF DIABETIC KETOACIDOSIS

Coexistent Lactic Acidosis. If the patient has coexistent lactic acidosis (shock), the qualitative serum test for ketones may not be strikingly positive (see above). Therefore, if hyperglycemia and glycosuria are present in the absence of or in the presence of only moderate ketonemia, suspect coexistent keto- and lactic acidosis. Enzymatic determinations for β-hydroxybutyrate and lactate confirm that diagnosis.

Absence of Increased Plasma AG. Under several circumstances, the patient with diabetic ketoacidosis does not have a large increase in the plasma AG.

Hypoalbuminemia. Hypoalbuminemia due to diabetic glomerulosclerosis is common in diabetes and obscures the increase in plasma AG (page 51). These patients have hyperglycemia, ketonemia, metabolic acidosis, ketonuria, and glycosuria, and you should not be deterred from the diagnosis by the normal or only slightly elevated level of the plasma AG if the serum ketones are positive to a dilution of 1:8.

Marked Ketonuria. If the patient has impaired proximal tubular reabsorption of ketones and glucose, the blood sugar may only be mildly elevated and the serum unmeasured anions not very increased in the presence of ketoacidosis. The clue to the diagnosis is metabolic acidosis and ketonuria. The diagnosis can be suspected by analyzing the urine electrolytes and osmolality (see page 54). To confirm the diagnosis, one should determine the urine β-hydroxybutyrate excretion rate.

the clinician should recognize the deficits present (Table 3–10, page 69). The therapeutic approach to diabetic ketoacidosis involves attention to 3 major issues:

1. *STOP KETOACID PRODUCTION* (see page 68). Regular insulin is the basis of the initial therapy for diabetic ketoacidosis. An initial bolus of 5–10 units is given intravenously, and a continuous infusion of 2–10 units/hr (in normal saline) is started. Even though the lipolytic rate declines promptly, there is a lag period of several hours before there is a net decline in ketoacids. Hence, expect little rise in the plasma bicarbonate concentration in the first several hours, despite insulin action. The plasma AG should return to normal in 8–10 hours, which reflects the course of the serum ketones.

2. *REEXPAND ECF VOLUME AND LOWER BLOOD GLUCOSE.* If the patient is in impending shock (systolic BP <90 mm Hg with tachycardia) and is ECF volume-depleted, use an ISOTONIC Na-containing solution at 1000 ml/30 min until systolic BP >100. If hemodynamically stable (systolic BP >100 mm Hg), use ½ isotonic Na solutions at 0.5–1 L/hr depending on degree of ECF volume depletion. If the patient is severely acidemic ([HCO_3] <5 mM), give some of the Na as $NaHCO_3$ until the plasma [HCO_3^-] >10 mM (see page 57).

Both insulin and saline contribute to the fall of the blood glucose concentration. With this therapy, the blood sugar usually falls at a rate of 100 mgm/dl/hr (5.5 mM/hr). Glucose should be added to the infusion once the blood sugar reaches 250 mgm/dl (12–15 mM).

3. *K STATUS* (see Table 3–10, page 69). There are 3 parameters to evaluate with respect to K replacement therapy: the plasma [K], the urine [K], and the urine flow rate. Patients with diabetic ketoacidosis usually are K depleted; therefore, if the plasma [K^+] is <5 mM and the urine output is brisk, and the urine [K] >20 mM, add 20–40 mmol KCl to each L of infusion. If plasma [K^+] <4 mM, use HCO_3^- with caution (as HCO_3^- causes K entry into the ICF).

If plasma [K^+] >5 mM, wait at least 1 hr before initiating K replacement therapy. However, if the plasma [K^+] >5 mM with the urine [K] >20 mM and the urine flow rate >2 ml/min, it is reasonable to add 20 mmol KCl/L to the intravenous solutions. In the first 6 hr, it is also reasonable to use some K phosphate (up to 6 mmol/hr), as these patients are also phosphate-depleted.

TABLE 3–9. Changes in $[HCO_3^-]$ and AG During Treatment of Diabetic Ketoacidosis

Condition	ECF Volume	ECF Concentration		ECF Content*		
		HCO₃	*KB*	*HCO₃*	*KB*	*SUM*†
	L	(mM)		(mmol)		
Normal	15	25	0	375	0	*375*
Admission	12‡	10	15	120	180	*300*‡
Early Rx§	15	12	5‖	180	75	*255*‖
Later Rx§	15	18	1	270	15	*285*¶
Recovery	15	25	0	375	0	*375*¶

*The ECF content is the ECF concentration × ECF volume.

†The sum of $[HCO_3^-]$ plus [KB] indicates the total "potential" HCO_3 content of the ECF (KB = ketone bodies).

‡The fall in ECF volume (osmotic diuresis) is highlighted—this reveals the ECF HCO_3^- deficit (content only 300), even though the rise in ketone body (KB) concentration equals the fall in $[HCO_3^-]$.

§Early therapy refers to the expected response in the first 4–6 hr after insulin action, whereas later therapy time would be 12–15 hr after this therapy.

‖Despite the conversion of ketone bodies to HCO_3, there is not an equivalent rise in ECF $[HCO_3]$. This reflects 3 processes: the infusion of HCO_3-free solution to expand the ECF volume, the exit of H^+ in excess of ketone anions from the ICF, where they were buffered on ICF proteins, and continued ketonuria with Na^+ and K^+.

¶With increased renal new HCO_3 generation, the HCO_3 deficit will be repaired.

TABLE 3–10. Deficits in Diabetic Ketoacidosis

Substance	Typical Deficit	Therapy
Na	5–10 mmol/kg	1–2 L isotonic saline 2+ L ½ isotonic saline
K	5–10 mmol/kg	KCl 20–40 mmol/L* Add phosphate 6 mmol/hr
Water	3 L from ECF 3 L from ICF	Give free water, especially if hypernatremic
HCO_3	2–3 mmol/kg	$NaHCO_3$ only if life-threatening acidosis. Later, only if low renal NH_4 excretion
Phosphate	0.5 mmol/kg	Deficit not life-threatening, but therapy advisable (6 mmol/hr)

*Owing to the osmotic diuresis, anticipate an ongoing renal excretion of at least 100 mmol Na. Probably half of the K administered will appear in the urine, despite the K deficit (mineralocorticoid- and flow-induced).

Lactic Acidosis

Pathogenesis. Lactic acidosis is a common cause of metabolic acidosis, often in a life-threatening situation. The commonest cause of lactic acidosis is relative hypoxia (O_2 demand exceeding the O_2 supply—Type A lactic acidosis). This may be due to hypoxemia, hypotension, or impaired blood supply to an organ. There are a number of causes of lactic acidosis in which hypoxia does not play a major role (Type B lactic acidosis). Almost all of these patients have liver problems that result in reduced lactate clearance. The biochemical basis of lactic acidosis is detailed on page 71.

- TYPE A, CAUSE = HYPOXIA: ACUTE PROBLEM
- TYPE B, CAUSE = LIVER PROBLEM: CAN BE CHRONIC
- TREATMENT DEPENDS ON CAUSE
  TYPE A: MUST STOP H^+ PRODUCTION ($\uparrow O_2$ TO TISSUES)

Quantitative Aspects (Table 3–11, page 71). In Type A lactic acidosis, the rate of H^+ production is close to 60 mmol/min with total body anoxia. Thus, survival for more than a few minutes depends entirely on stopping H^+ production. However, hypoxia rather than anoxia is usually the case. If some hypoxia is present and it involves the liver, life-threatening lactic acidosis is produced. Hence, therapy must be directed at normalizing liver O_2 supply.

Diagnosis. The diagnosis of Type A lactic acidosis must be made quickly, owing to the high rate of H^+ production (Table 3–11); it is easily established on the basis of the clinical setting of low O_2 content (severe anemia or cyanosis) or very poor O_2 delivery to tissues (shock) (Table 3–12, page 73). There may also be symptoms of ischemia in one region of the body. However, for severe lactic acidosis, the patient must also be hypotensive because:

1. With local arterial inflow obstruction, the reduction in O_2 delivery is accompanied by a reduction in the delivery of the lactic acid precursor, glucose.
2. A normally perfused liver metabolizes the lactic acid produced, yielding glucose.

In Type B lactic acidosis, the diagnosis may be less

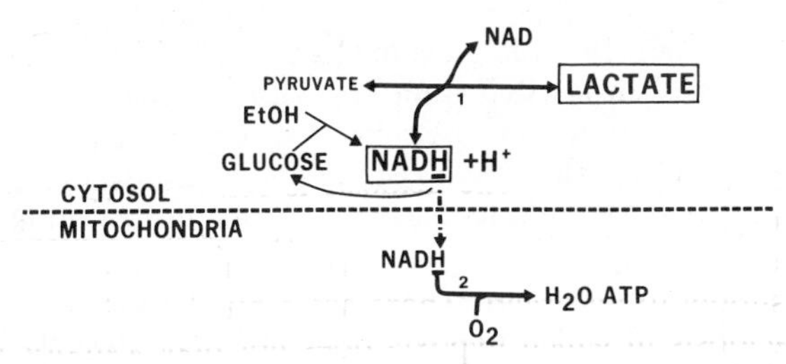

Figure 3–5. Biochemistry of lactic acidosis. Reactions above the dashed line are cytoplasmic and below are mitochondrial. Lactate is formed by only one reaction in the cytoplasm of all cells. The enzyme catalyzing this near-equilibrium step (regulated only by the concentrations of substrates and products) is lactate dehydrogenase, Reaction 1. Thus, lactate is a metabolic "dead end" that can accumulate only if NADH or pyruvate accumulate. This provides the basis for the classification in Table 3–12, page 73. NADH or pyruvate accumulate when their formation rates exceed their utilization rates.

TABLE 3–11. Duration Required to Titrate 50% of Body Buffers During Hypoxia

% Fall in O_2 Utilization	H^+ Production (mmol/min)*	Time to Titrate 470 mmol Buffer† (min)‡
100	62	8
50	31	16
25	16	32
10	6	80
5	3	160

*The oxidation of 450 g of glucose (2500 mmol) to CO_2 could supply the entire daily ATP requirement (1800 KCal/day). However, if this ATP were supplied by glycolysis, 90,000 mmol of lactic acid would be produced (2500 mmol × 36 ATP/glucose; glycolysis yields 2 ATP/2 lactic acids). Since there are 1440 min/day, with 100% anoxia 62 mmol of lactic acid would be produced/min.

†Total physiologic buffer capacity is close to 940 mmol; this represents half of the total body buffers.

‡Should the liver not be hypoxic, most of the lactic acid produced is converted back to glucose. This greatly prolongs the survival time. Hence, hepatic hypoxia plays a central role in producing life-threatening lactic acidosis.

obvious. There is either evidence of a liver problem or of the intake of drugs that interfere with hepatic metabolism. Alternatively, there may be signs of a very large tumor load. In addition, the other causes of metabolic acidosis associated with an increase in the plasma AG should be ruled out (Table 3–1, page 45).

The diagnosis of Types A and B lactic acidosis can be confirmed with an enzymatic determination of the plasma L-lactate concentration.

Treatment

> IN TYPE A LACTIC ACIDOSIS, YOU MUST LOWER THE H^+ PRODUCTION RATE

Type A Lactic Acidosis. The only effective treatment is to correct the reduced O_2 supply.

1. Restore blood pressure in hypotension with Na- and albumin-containing solutions, as dictated by the clinical circumstances (often circulating volume contraction). Improve myocardial function and thereby tissue perfusion.
2. Correct hypoxemia with O_2.
3. Correct severe anemia with transfusion.
4. Resect ischemic area if necrotic; if not, restore its blood supply.
5. Treat sepsis (it decreases O_2 delivery and interferes with O_2 extraction). Expand the circulating volume.

$NaHCO_3$ Therapy in Type A Lactic Acidosis

The use of $NaHCO_3$ in total anoxia is of little value, owing to the magnitude of the H^+ load (Table 3–11). Nevertheless, in cases in which the hypoxia is marginal but potentially reversible, $NaHCO_3$ may buy time by improving myocardial function, but even this is controversial. The Na^+ load may pose a major limit to this type of therapy. If this is anticipated, use a loop diuretic or consider dialysis therapy against a HCO_3 bath if technically feasible.

The other measure that can lead to endogenous HCO_3 formation is the use of the activator of pyruvate dehydrogenase, dichloroacetate. Recall that this metabolism requires O_2 consumption and is of limited value in these patients, unless it helps restore O_2 delivery by improving myocardial function (there is some evidence that this occurs). Again,

TABLE 3–12. Classification of Lactic Acidosis

REDUCED O_2 DELIVERY VS DEMAND (TYPE A)

Low Blood Oxygen Content

- Low arterial Po_2 (o_2-poor air or lung problem)
- Hemoglobin problem (severe anemia, altered hemoglobin)

Decreased Tissue Perfusion

- Local arterial obstruction
- Loss of arterial blood volume (hemorrhage, venous pooling as in septic shock)
- Cardiac disease

High Tissue Demand for O_2

- Severe exercise
- Uncouplers of oxidative phosphorylation
- Hypermetabolism (fever, thyrotoxicosis)
- Very large tumor burden

REDUCED LACTATE REMOVAL WITHOUT HYPOXIA (TYPE B)

Liver Disease (extensive)

- Infiltration by tumor cells
- Damage (e.g., cirrhosis)

Interference With Normal Liver Metabolism (Fig. 3–5)

- Increased NADH production (ethanol)
- Decreased NADH oxidation (cyanide)
- Decreased pyruvate metabolism (e.g., drugs such as phenformin; inborn errors of metabolism such as hereditary fructose intolerance or gluconeogenic enzyme defects; nutritional problems such as thiamine deficiency)
- Increased pyruvate production (e.g., very large tumor burden)

TABLE 3–13. Substances Associated with Lactic Acidosis

Drugs	Alcohols	Sugars
Biguanides	Ethanol	Fructose
Streptozotocin	Methanol	Sorbitol
Salicylate		Xylitol
Isoniazid		

this is a temporary measure, as excessive consumption of glucose will be the net result of this therapy.

Type B Lactic Acidosis. There is not the same urgency here, since the rate of H^+ accumulation is much lower than in Type A lactic acidosis.

1. *Lactic acidosis due to thiamine deficiency:* This form of lactic acidosis is entirely preventable by giving thiamine. The clinical setting in which it is most prevalent is in the alcoholic or in patients in whom nutritional intake is inadequate. The major danger is not the acidosis but rather the CNS lesion that is caused by local glycolysis accelerated by reversal of alcoholic ketoacidosis with glucose therapy.
2. *Ethanol-induced lactic acidosis (Fig. 3–5, page 71):* The cause of lactic acidosis in this case is hepatic ethanol metabolism generating NADH, leading to diversion of pyruvate to lactate. Thus ethanol metabolism must be present when you make this diagnosis. Furthermore, since all other tissues can oxidize the lactate produced, the degree of lactic acidosis is always mild if this is the sole cause. No specific treatment is required.
3. *Drug-induced lactic acidosis (Table 3–13, page 73):* The drugs in question accumulate if intake is high or excretion low (especially prerenal failure in the case of use of phenformin). Although the degree of lactic acidosis may be severe, chances of survival are good. The measures required are to neutralize the H^+ excess with $NaHCO_3$, slow H^+ production with insulin in the case of phenformin-induced lactic acidosis, accelerate lactate metabolism (dichloroacetate), and eliminate the drug by renal excretion or dialysis.
4. *Other forms of lactic acidosis (see page 75).*

ORGANIC ACID LOAD FROM THE GI TRACT (D-LACTIC ACIDOSIS)

Certain bacteria in the GI tract may convert carbohydrate (cellulose) into organic acids. The two factors that make this possible are slow GI transit (blind loops, obstruction) and change of the normal flora (usually with antibiotic therapy). The most prevalent organic acid is D-lactic acid. Since humans metabolize this isomer more slowly than L-lactate and production rates can be very rapid, life-threatening acidosis can be produced.

There are 3 additional points that should be noted with respect to D-lactic acidosis:

Other Causes of Type B Lactic Acidosis

Overproduction of Lactic Acid. Lactic acid can be produced at a rate exceeding hepatic clearance during exercise. This results in a temporary physiologic lactic acidosis and has no clinical importance in almost every case. However, the degree of lactic acidosis may be up to 15 mM. An exaggerated form of this type of lactic acidosis is seen in patients during convulsions.

Lactic Acidosis Due to Tumors. There are 3 major subgroups to consider:

1. Tumor infiltration of the liver, which compromises the normal hepatic lactate clearance.
2. The release of metabolites from necrotic tumor masses can lead to inhibition of hepatic gluconeogenesis (e.g., a product of the amino acid tryptophan can inhibit phosphoenolpyruvate carboxykinase, a key gluconeogenic enzyme).
3. It is possible that exceedingly large tumor burdens can synthesize so much lactic acid, given their biochemistry, that it would exceed the capacity of even a normal liver to remove it.

Obviously, all 3 mechanisms may operate to produce this type of lactic acidosis. In addition, should hypoxia or hypotension supervene, the lactic acidosis would be greatly aggravated.

Treatment must be directed at the primary causes, and as expected, prognosis generally reflects the behavior of the tumor and the liver disease.

Hepatic Insufficiency or Inborn Errors of Hepatic Metabolism. These are not discussed here. Survival depends on treating the underlying cause. In the case of inborn errors, sometimes specific therapies are possible (avoiding fructose in hereditary fructose intolerance or modest glucose administration in glycogen storage disease, Type 1, with hypoglycemia, and vitamin therapy in certain enzymatic defects).

POTENTIAL DISADVANTAGES OF $NAHCO_3$ THERAPY IN CHRONIC LACTIC ACIDOSIS

Lactic acid production rate declines as the $[H^+]$ rises. Therefore, the administration of $NaHCO_3$, by diminishing the $[H^+]$, increases glucose conversion to lactic acid. This may have 2 important clinical implications:

1. If the source of the glucose is the lean body mass, considerable muscle mass may be lost as a result of $NaHCO_3$ therapy.
2. If the $NaHCO_3$ is given in glucose and water (isotonic), 600 mmol of H^+ could be produced from the glucose, whereas only 150 mmol of HCO_3^- were provided.

1. The usual laboratory test for lactate is specific for the L-lactate isomer. Hence, the laboratory report for "lactate" is not elevated. To confirm this diagnosis, measure the plasma D-lactate concentration.

2. GI bacteria produce amines and other compounds that may cause clinical symptoms related to CNS dysfunction (personality changes, gait changes, confusion, etc.).

3. Some of the D-lactate is lost in the urine if the GFR is not too low. Hence, the degree of rise in the plasma AG may not be as high as expected for the fall in the plasma [HCO_3^-].

Treatment is directed at the gastrointestinal problem and ensuring that the patients do not die of severe metabolic acidosis (give $NaHCO_3$ if necessary).

Questions (Answers on page 77)

12. What is the significance of the anion that accompanied the acid during metabolic acidosis?

13. Why is the metabolic acidosis of renal failure associated with a wide plasma AG?

Renal Failure (see Fig. 3–6 and Answer 13, page 77)

- Requires GFR <20% of normal to raise the plasma AG significantly. For each 1 mg/dl increase in serum creatinine, the plasma AG increases by 1 mEq/L (a rough guide).
- Requires low NH_4^+ excretion to lower plasma [HCO_3^-]. Degree of lowering of [HCO_3] depends on magnitude of dietary acid (protein) load vs. NH_4^+ excretion rate.
- Added danger of hyperkalemia.
- Treatment: see page 58 for acute therapy. Consider long term management (e.g., dialysis).

Metabolic Acidosis Due to Intoxications

Wide Plasma Osmolal Gap Types (see Answer 14, page 85)

PLASMA OSMOLAL GAP = MEASURED OSMOLALITY − CALCULATED OSMOLALITY (2 × [Na] + [Glucose]/18 + [Urea]/2.8, both the latter in mg/dl)

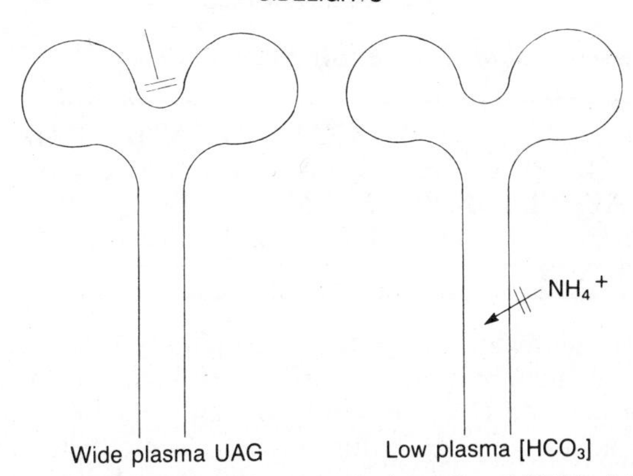

Figure 3–6. Independent effects on plasma AG and $[HCO_3^-]$ in renal failure. The rise in the plasma AG in renal failure is due to a low GFR. This is depicted in the schematic of a nephron on the left. The fall in the plasma $[HCO_3^-]$ in renal failure is due to a low rate of NH_4^+ excretion secondary to reduced NH_4^+ production. This is primarily a tubular problem, as depicted in the schematic on the right. Depending on the specific nature of the renal diseae, either process may be affected to a greater degree than the other.

Answers

12. Although the H^+ merits most of the clinical attention, the anion is significant because:

It is a useful clue (via the plasma AG) to detect the presence of metabolic acidosis and aids in its differential diagnosis.

With an organic acidosis (e.g., lactic acidosis), when the abnormality is reversed, each mmol of lactate will yield a mmol of HCO_3^-.

It is important that there be little renal loss of this anion with Na or K because this is equivalent to the loss of potential $NaHCO_3$. However, there is no acid-base impact if the anion is excreted along with H^+ or NH_4^+.

13. This question should be answered in 2 parts (Fig. 3–6):

1. The basis of the metabolic acidosis is best thought of in terms of HCO_3 balance. These patients have the usual ongoing HCO_3 consumption (1 mmol/kg/day) due to the dietary H^+ production. They also have impaired renal *new* HCO_3 generation (reduced NH_4^+ excretion). Thus, there is continued net HCO_3^- consumption, leading to a slowly progressive metabolic acidosis.

2. Certain unmeasured anions are excreted via filtration. If the GFR falls to ½ of normal, the plasma level of these anions has to double to achieve the same filtered load and thereby excretion rate. Thus, patients with a low GFR develop an increase in their plasma AG. Although phosphate and sulphate accumulate in renal failure, they do not account for all of the rise in the plasma AG.

Methyl Alcohol (Methanol) Intoxication

- SUSPECT IF: INTOXICATED PATIENT HAS A WIDE PLASMA AG *AND* A WIDE OSMOLAL GAP; VISION PROBLEM.
- TREATMENT: ETHANOL, $NaHCO_3$ AND DIALYSIS.

When methanol is ingested, the quantity is so large that it can, by metabolism, lead to a very high rate of formic acid production (1 oz of methanol can yield 1000 mmol of formic acid—see page 79 for details). As methanol must be metabolized via hepatic alcohol dehydrogenase, you can expect a H^+ production rate of up to 300 mmol/hr due directly to methanol metabolism, if blood levels are high.

Since methanol is widely available and inexpensive, it is an intoxicant that must always be considered in the differential diagnosis of metabolic acidosis with a wide plasma AG. Furthermore, since a product of methanol metabolism (probably formaldehyde) is a major neurotoxin that has special predilection for the optic nerve, it is especially important to prevent this complication.

The clues to suspect the diagnosis are: a history of ingestion, the sweet smell of methanol on the breath, the presence of a wide osmolal gap in the plasma (see Answer 14, page 85); later, confirm the diagnosis by the finding of high blood methanol levels.

Treatment. There are 3 major components of therapy:

1. Establishment and maintenance of adequate ethanol levels to prevent methanol metabolism.
2. Correction of life-threatening acidemia.
3. Removal of methanol.

Ethanol Therapy (see page 79 for discussion). (1) *Establish therapeutic blood level of ethanol* of 100 mg/dl (22 mM) by giving 0.6 g of ethanol per kg body weight, IV or orally (4 oz whiskey orally). (2) *Maintain blood ethanol level for duration of intoxication,* i.e., in chronic drinkers, give 0.15 g/kg/hr; in nondrinkers give 0.07 g/kg/hr orally or IV (1–2 oz whiskey po/hr).

Correction of Acidemia (see page 54 for approach).

Removal of Methanol. Methanol is best removed by hemodialysis, but if unavailable, peritoneal dialysis can remove some methanol. Blood ethanol levels can be maintained by adding ethanol to the hemodialysate.

PATHOGENESIS OF METHANOL INTOXICATION AND RATIONALE FOR THERAPY

The toxicity of methanol and the associated severe acidemia are both related to the metabolism of methanol to formaldehyde and the organic acid, formic acid, catalyzed initially by hepatic alcohol dehydrogenase. As methanol itself is relatively nontoxic, the approach to therapy is to retard its metabolism by competing for availability of *alcohol dehydrogenase* and then to remove the methanol by dialysis. Since alcohol dehydrogenase also catalyzes the metabolism of ethanol, and because its affinity for ethanol is far greater than for methanol, the provision of ethanol is an effective way to lower the rate of methanol metabolism.

Chronic drinkers may have higher levels of alcohol dehydrogenase; therefore, the chronic alcoholic could be more threatened by methanol intoxication (by both acidemia and toxicity). However, if the alcoholic has also been drinking ethanol recently, and then is exposed to methanol, he is more protected than others who have not coincidentally ingested ethanol. For the same reason (higher alcohol dehydrogenase levels), the chronic drinker requires a higher rate of ethanol administration to maintain therapeutic levels of ethanol (see page 78 for detailed treatment protocol).

The blood level of ethanol should be monitored during treatment, and the dosage used may require modification, depending on individual differences in rates of absorption, metabolism, and removal of ethanol.

Methanol Conversion to Formic Acid

1. Assume 1 oz of methanol = 32 g
2. The molecular weight of methanol = 32

1 oz methanol = 1000 mmol (32000 mg/32)

3. Maximum activity of hepatic alcohol dehydrogenase. The liver can metabolize ethanol at a maximum rate of 300–400 g/day (close to 7200 mmol of ethanol/day). This is equivalent to a maximum rate of 300 mmol/hr. Hence, methanol metabolism can produce up to 300 mmol H^+/hr, assuming the same V_{max} for methanol.

Ethylene Glycol (Antifreeze) Intoxication

SUSPECT IF:
- INTOXICATED, WITH METABOLIC ACIDOSIS
- WIDE PLASMA AG AND OSMOLAL GAP
- URINE OXALATE CRYSTALS
- ACUTE TUBULAR NECROSIS

TREATMENT:
- ETHANOL
- $NaHCO_3$
- DIALYSIS

Because ethylene glycol is readily available, relatively inexpensive, and pleasant tasting, it may be ingested as an intoxicant. It causes fulminant metabolic acidosis, severe central nervous system toxicity, and acute tubular necrosis. As with methanol, the products of ethylene glycol metabolism cause the toxicity (page 81). Following the initial toxicity associated with profound metabolic acidosis and CNS manifestations (confusion, coma, seizures), patients may develop congestive heart failure and respiratory failure. Those who survive usually have acute tubular necrosis that is generally of the oliguric form. Ethylene glycol intoxication should always be suspected in patients with metabolic acidosis and an increased plasma AG, especially if the patient appears intoxicated and denies ethanol intake or if the odor of ethanol is not evident. The index of suspicion can be increased greatly by finding oxalate crystals in the urine. As with methanol, a wide plasma OSMOLAL GAP is helpful in the absence of ethanol (see Question 14). The diagnosis is confirmed by detecting ethylene glycol in the blood.

Therapy. The principles of therapy of ethylene glycol intoxication are identical to those for methanol. The additional complication of acute tubular necrosis may limit the quantity of $NaHCO_3$ that can be given, as this form of intoxication is generally oliguric. Therefore, acute pulmonary edema may become a feature in the therapy for ethylene glycol intoxication. The use of nasogastric suction and pentagastrin stimulation of gastric acid secretion may ameliorate both the acidemia and the pulmonary edema while dialysis is being arranged (see page 57). Because of the acute renal failure, early dialysis is critical. As with methanol, hemodialysis is the preferred manner to remove

TABLE 3–14. Products of Metabolism of Ethanol, Methanol, and Ethylene Glycol

Alcohol	ADH* ⟶	Aldehyde	⟶	Carboxylic Acid
Ethanol	→	Acetaldehyde	→	Acetic Acid
Methanol	→	Formaldehyde†	→	Formic Acid
Ethylene glycol‡	→	Glyoxalic Acid§	→	Oxalic Acid‖

*ADH = hepatic alcohol dehydrogenase. Ethanol is the preferred substrate and therefore by competition prevents the metabolism of methanol and ethylene glycol.

†Formaldehyde is probably the neurotoxin. Therefore, prevent methanol metabolism by giving ethanol, and then remove methanol by dialysis (ensure ethanol added to dialysate or supplemented IV).

‡Ethylene glycol is not really an alcohol, but it is a "dialcohol" (2 adjacent carbons, each having a hydroxyl group).

§Glyoxalic acid accumulates and circulates.

‖Oxalic acid precipitates in solutions containing calcium and may be the basis of the high incidence of acute tubular necrosis upon metabolism of ethylene glycol. Therapy with ethanol and dialysis reduces the glyoxalic acid production.

KINETICS OF METHANOL AND ETHYLENE GLYCOL METABOLISM

The affinity of methanol and ethylene glycol for alcohol dehydrogenase is much lower than that of ethanol (i.e., 100-fold higher K_m). Therefore:

1. If the blood concentration of the alcohol is at least twice its K_m, the enzyme proceeds at close to its maximum velocity.
2. As the blood concentration falls somewhat below the K_m, the rate of removal of that substance diminishes appreciably.
3. Since the K_m of ethanol is 100-fold lower than that of either methanol or ethylene glycol, lower levels of ethanol prevent the metabolism of higher concentrations of methanol or ethylene glycol.
4. In clinical practice, methanol and ethylene glycol intoxication are characterized by high rates of H^+ production at high blood levels and a slow fall of the elevated blood levels from 40 mM on down.
5. At low levels of intoxication, the metabolism is so slow that the intoxicant can be excreted in the urine without undergoing metabolism.

ethylene glycol; however, if it is not available, peritoneal dialysis should be carried out. Again, ethanol levels should be maintained during hemodialysis by infusion or addition to the bath.

Normal Plasma Osmolal Gap Types

(a) Salicylate Intoxication (see page 83)

- USUALLY RESPIRATORY ALKALOSIS
- METABOLIC ACIDOSIS IN CHILDREN
- PROBLEM IS TISSUE ASA LEVELS
- TREATMENT:
  - PROMOTE URINE EXCRETION WITH ALKALINE DIURESIS
  - AVOID ACIDEMIA AND SEVERE ALKALEMIA IN THERAPY

Although salicylate intoxication is very common, it only causes metabolic acidosis in children; the younger the child, the more likely the metabolic acidosis. The most common acid-base disturbance associated with salicylate intoxication is RESPIRATORY ALKALOSIS due to central respiratory stimulation. Acid-base disturbances are the prominent feature of acute, rather than chronic, salicylate intoxication. As you can readily appreciate, since the toxic level of salicylate is 3–5 mM, the plasma unmeasured anions associated with the metabolic acidosis are not only salicylate, but rather are ketone bodies, sometimes lactate and other unidentified organic acid anions.

Diagnosis. The diagnosis of salicylate intoxication might be suspected from the history of ingestion or the symptoms of tinnitus and lightheadedness and the presence of a respiratory alkalosis complicating the metabolic acidosis. The suspicion is increased by finding unexplained ketosis (salicylate activates tissue lipase), hypouricemia (high-dose salicylate is uricosuric) and a urine (Na + K) >>Cl. It is confirmed by detecting salicylates in the blood.

Treatment. Generally, metabolic acidosis is not a serious feature of salicylate intoxication. In the absence of severe metabolic acidosis, the therapeutic efforts in salicylate intoxication are to promote salicylate excretion. The maneuvers include:

ALKALI THERAPY IN SALICYLATE INTOXICATION (Figs. 3–7A and 7B)

If the patient with salicylate intoxication has metabolic acidosis, acidemia should be corrected because it increases the concentration of nonionized salicylic acid (HASA) in the blood. Since this uncharged form crosses cell membranes, its diffusion into the CNS cells is facilitated, promoting toxicity.

In an analogous manner, an alkaline urine pH promotes salicylate excretion, converting HASA to the salicylate (ASA^-) anion, thereby retarding HASA reabsorption by nonionic diffusion. The problem with aggressive HCO_3^- therapy is that the patient with respiratory alkalosis may become very alkalemic.

$H^+ + ASA^- \rightleftharpoons HASA$
ONLY HASA CROSSES CELL MEMBRANES.

Acetazolamide and ASA Excretion. Acetazolamide can be used in therapy of ASA intoxication. Its major value is to cause HCO_3^- excretion in patients given $NaHCO_3$ who are severely alkalemic. Potential disadvantages are metabolic acidosis causing more ASA toxicity (see above) and a lowering of proximal tubular fluid pH, negating some of the benefits of the HCO_3 diuresis (final urine pH is more alkaline after the spontaneous dehydration of H_2CO_3, but the $[H_2CO_3]$ in the proximal tubule is increased by acetazolamide.

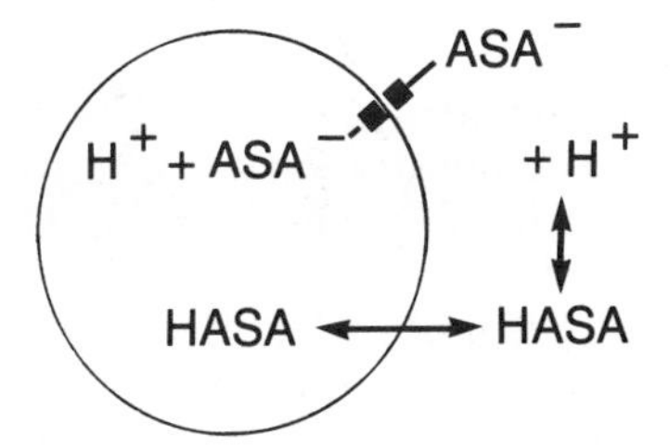

Figure 3–7A. Role of alkali in salicylate intoxication: **Serum.** Alkali therapy increases the ionized [salicylate⁻], which reduces diffusion into the ICF (see preceding text for details).

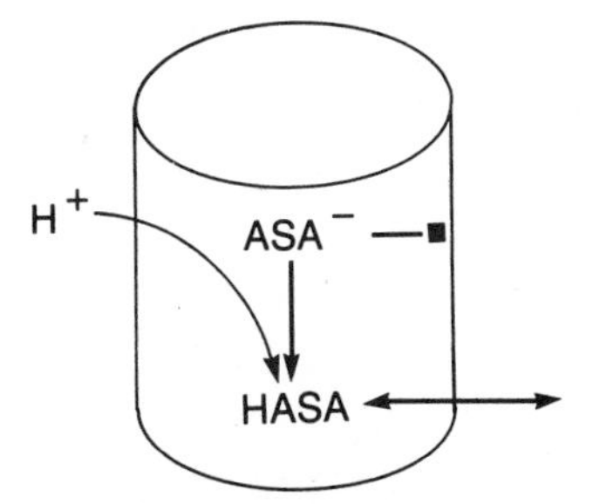

Figure 3–7B. Role of alkali in salicylate intoxication: **Urine.** Alkalinization of the urine prevents the formation of HASA by secreted H^+, leaving the ASA^-, which is relatively nonreabsorbable, thus increasing the renal excretion of salicylate.

Alkaline Diuresis. Caution must be exercised in view of respiratory alkalosis (page 83). Infusion of ½ N saline containing 50 mmol $NaHCO_3$/L at a rate of 300–500 ml/hr produces an alkaline diuresis. If the patient cannot tolerate that Na load, a loop diuretic may be used to promote Na excretion. The blood pH should be monitored hourly, and if it exceeds 7.55, 500 mg acetazolamide should be given to promote HCO_3^- excretion.

2. In severe intoxications, hemodialysis should be used, or, in its absence, peritoneal dialysis.

Question (Answer on page 85)

14. Patients A, B, and C listed below each have metabolic acidosis and a wide plasma AG. Which one has methanol intoxication, renal failure, or D-lactic acidosis?

Patient	A	B	C
Calculated Osmolality	290	290	320
2 × plasma [Na]	280	280	280
BUN (mg/dl/2.8)	5	5	35
Glucose (mg/dl/18)	5	5	5
Measured Osmolality	290	320	320

PARALDEHYDE

Paraldehyde was widely used in the past because of ease of administration and effectiveness. This agent is now rarely used. Metabolism of paraldehyde leads to metabolic acidosis.

Diagnosis. The diagnosis of paraldehyde intoxication is not difficult, as this agent is excreted via the lungs with a characteristic pungent fetor on the breath.

Therapy. Treatment with alkali is usually sufficient to deal with paraldehyde intoxication.

Answers

14. This question emphasizes the diagnostic importance of the calculation of the plasma osmolality from 2 times the [Na] plus the [Glucose] and the [Urea] and the comparison of this calculated osmolality with the measured osmolality (the plasma osmolal gap). If one is dealing with SI units, the mM values also are the osmotic contribution of the substance, whereas if the concentrations are in mg/dl, one must multiply by 10 (to obtain mg/L and divide by the molecular weight [BUN 28, glucose 180] to obtain mmol/L). The molecular weights of methanol, ethanol, and ethylene glycol are 32, 46, and 62, respectively.

In methanol intoxication (as with ethanol or ethylene glycol intoxication), the intoxicant is of high enough concentration so that it significantly raises the plasma osmolality. This increase is evident when the measured osmolality is compared with that calculated from the values for Na, urea, and glucose concentrations.

Case B has a large osmolal gap (calculated 290 mOsm/kg, measured 320 mOsm/kg) and therefore represents a patient with a circulating osmole that is neither Na nor its anions (the reason the [Na] is multiplied by 2) nor urea nor glucose—i.e., the patient with methanol intoxication.

Case C also has an increased measured osmolality, but in this case it is due to the urea, and this is the patient with renal failure. This case emphasizes the importance of considering the contribution of urea whenever using the osmolality.

Case A represents the patient with D-lactic acidosis, since in this case, the serum osmolality is in the normal range—not elevated either by methanol or by urea.

TABLE 3–15. Metabolic Acidosis with No Increase in Plasma AG

Associated With Increased Renal NH_4^+ Excretion
GI HCO_3^- Loss
Acid Ingestion (HCl)
Proximal RTA (when urine pH <6.0)
After Acetazolamide Ingestion
After Hypocapnia
Expansion Acidosis
Ketoacidosis and Marked Ketonuria
Associated With Low Renal NH_4^+ Excretion
Reduced NH_3 Availability
Decreased Ammoniagenesis
• Low GFR
• Hyperkalemia
Medullary Interstitial Disease
Reduced Collecting Duct H^+ Secretion
H^+ Pump Failure
Impaired H^+ Pump Regulation
H^+ Backleak (lumen to cell)

Metabolic Acidosis and No Increase in the Plasma AG (Table 3–15, page 85)

The diagnosis of metabolic acidosis and no increase in the plasma AG can best be pursued by dividing the patients into 2 broad groups based on urine NH_4^+ excretion.

- ASSOCIATED WITH NORMAL RENAL NH_4^+ EXCRETION
- ASSOCIATED WITH REDUCED RENAL NH_4^+ EXCRETION

With Normal Renal NH_4^+ Excretion

These patients are generally identified by the urine apparent net charge, which is negative (urine $[Cl^-] > ([Na^+] + [K^+])$, indicating the presence of an unmeasured cation (NH_4^+) (see page 54). These patients have a "nonrenal" basis for their metabolic acidosis, since renal NH_4^+ excretion should be sufficient to prevent metabolic acidosis. The commonest cause of this disorder is GI HCO_3^- loss.

Gastrointestinal HCO_3^- Loss. Diarrhea is easily diagnosed; however, HCO_3^--rich fluid in the bowel lumen (ileus) may not be so evident. Nevertheless, the patient with GI HCO_3^- loss has a normal renal response to acidemia (a high urine NH_4^+ excretion rate revealed by the urine "apparent net charge," when urine $[NH_4^+]$ is substantial, the urine [Cl] is > [Na] plus [K]). These patients often have hypokalemia and ECF volume depletion and therefore their condition could be confused with distal RTA.

Question (Answer on page 87):

15. An 80 year old man, with a history of "pyelonephritis," developed diarrhea after a course of antibiotics. On the basis of the following results, a diagnosis of distal RTA was made. Is it correct?

	Plasma		Urine
Na	134 mM	Na	10 mM
K	2.8 mM	K	40 mM
Cl	115 mM	Cl	100 mM
HCO_3	10 mM	OSM	800 mOsm/kg H_2O
H^+	62 nM (pH 7.20)	pH	5.9

Technique to Diagnose pRTA

When the patient is acidemic, administer $NaHCO_3$ and monitor the plasma [HCO_3] and urine pH. If the urine pH becomes alkaline (>7.0) while the plasma [HCO_3] remains low, impaired proximal H^+ secretion is present. With continued HCO_3 administration, the urine pH rises further, and distal H^+ secretion can be assessed by measuring the urine P_{CO_2} (see page 55).

Urine pH, [HCO_3^-], and [NH_4^+] in pRTA

1. Since patients are in steady state, 24 hr net acid excretion in pRTA is the same as in normals—i.e., HCO_3 gain = HCO_3 loss but at a lower plasma [HCO_3^-].
2. Variations occur during the day: (A) Gastric H^+ secretion results in a temporary rise in the plasma [HCO_3^+] above the low threshold. This causes bicarbonaturia and a rise in urine pH. (B) Pancreatic $NaHCO_3$ secretion results in a temporary further fall in the plasma [HCO_3^-]. This causes a rise in urine NH_4^+ excretion together with a fall in urine [HCO_3] and pH.
3. Therefore, urine values in pRTA must be interpreted relative to the clinical situation at the time of assessment.

TABLE 3–16. Diagnostic Features in Proximal RTA (pRTA)

State	Filtered HCO_3	Proximal HCO_3 Reabsorbed	Distal HCO_3 Delivery	Net HCO_3 Excretion
	(mmol/hr)			
Normal	180	150	30	−3*
Onset pRTA	180	90†	90	18‡
Established pRTA	150	90	60	−3
+ Acid load pRTA	120	90	30	−12§

*Normals excrete 70 mmol of net acid/day or 3 mmol/hr. This results in a HCO_3 gain for the body to titrate the daily H^+ (indicated by minus sign).
†The lesion in pRTA is reduced proximal H^+ secretion.
‡Since distal H^+ secretion is of low capacity, excess HCO_3 delivered is excreted, and the urine pH is >7.0.
§If an extra acid load is present, NH_4^+ excretion can rise, augmenting *new* HCO_3 formation.

Answers

15. Although the blood tests (metabolic acidosis with a normal plasma AG, and hypokalemia) and the history of "pyelonephritis" suggest the diagnosis of distal RTA, the urine electrolytes indicate that this diagnosis is incorrect. The [Cl] exceeds [Na] + [K], indicating significant NH_4^+ excretion that would not be present in distal RTA. Therefore, this patient's metabolic acidosis is primarily due to diarrhea, which also accounts for the hypokalemia (there may be some K loss in the urine, owing to the ECF volume contraction dependent on urine flow rate—see Chapter 8).

Acid Ingestion. If the anion of an acid is Cl, its intake can cause metabolic acidosis with no increase in the plasma AG. HCl, NH_4Cl, lysine, and arginine HCl cause this disorder. The diagnostic features are the same as with GI HCO_3 loss.

Failure to Recycle Filtered HCO_3^- (Proximal RTA, see also Table 3–16, page 87). Recycling of filtered HCO_3^- is achieved by proximal tubular H^+ secretion. A defect in this H^+ secretion results in a metabolic acidosis with no increase in the plasma AG and is due to the excretion of $NaHCO_3$ in the urine (called proximal renal tubular acidosis—pRTA, or Type II RTA). The H^+ secretory defect may be isolated, or may occur in concert with other transport defects, leading to the diagnosis of Fanconi syndrome. The diagnosis of pRTA involves demonstrating impaired reabsorption of filtered HCO_3^- (see page 87).

Other Causes With Normal Renal NH_4^+ Excretion

1. After acetazolamide use (Question 17).
2. Recovery from chronic hypocapnia.
3. Expansion acidosis (page 89).

Questions (Answers on page 89)

16. What are diagnostic features of disorders with reduced recycling of filtered HCO_3^-?
17. In what way will the urine findings change once acetazolamide is no longer acting but the patient still has metabolic acidosis?

WITH REDUCED RENAL NH_4 EXCRETION-DISTAL RTA (Table 3–18, page 91)

RENAL NEW HCO_3^- FORMATION MAY BE LOW BECAUSE OF:

1. LOW NH_3 AVAILABILITY TO THE COLLECTING DUCT.
2. LOW COLLECTING DUCT H^+ SECRETION.

When approaching the differential diagnosis of the patient with impaired renal *new* HCO_3 formation or low NH_4^+ excretion (also called Type I, classic, or distal RTA), it is helpful to think of 2 *major groups* based on the urine pH to determine why urine NH_4^+ is low (Fig. 3–3, page 53):

TABLE 3–17. Conditions Leading to Decreased Bicarbonate Recycling in the Proximal Convoluted Tubule

Proximal Renal Tubular Acidosis*
- Genetic disorders, including cystinosis, galactosemia, hereditary fructose intolerance, Wilson's disease, Lowe's syndrome, tyrosinemia
- Toxin-induced
  Exogenous toxins, including heavy metals, outdated tetracycline, streptozotocin
  Endogenous toxins, dysproteinemias, including multiple myeloma, etc.
- Others, including sporadic, transient proximal RTA of infants
- Secondary to other renal diseases, including amyloidosis, renal transplantation, autoimmune diseases such as chronic active hepatitis, Sjögren's syndrome, etc.

Carbonic Anhydrase Inhibitors
Others (Mild, e.g., Hyperparathyroidism, Hypocalcemia, Vitamin D Deficiency)

*This disorder may be specific for HCO_3 reabsorption or may be part of a generalized proximal tubular dysfunction.

Answers

16. The diagnostic features are an ALKALINE URINE at a time when PLASMA $[HCO_3^-]$ and pH ARE REDUCED.

17. Once the effects of acetazolamide have worn off, proximal HCO_3 recycling becomes normal and bicarbonaturia disappears. The patient has probably been acidemic for several days, so the urine $[NH_4^+]$ is high (urine [Cl] > [Na] + [K]).

Expansion Acidosis

When the ECF volume is expanded with solutions lacking HCO_3 (normal saline), the plasma $[HCO_3^-]$ falls. One must distinguish between HCO_3^- *content* and HCO_3^- *concentration* in the ECF.

In the table below, the patient had a normal plasma $[HCO_3^-]$ despite severe ECF volume contraction; thus, the ECF HCO_3^-*CONTENT* was reduced (line 2). Reexpansion of the ECF volume with normal saline (a bicarbonate-free solution) made the reduced HCO_3^- content evident as the $[HCO_3]$ fell (line 3). The previous loss of HCO_3^- was obscured by ECF volume contraction. For the $[HCO_3^-]$ to return to normal in this patient, he or she must either make new HCO_3 (excrete NH_4^+) or you must give $NaHCO_3$.

Condition	ECFV	$[HCO_3^-]$	HCO_3^- Content
Normal	15 L	24 mM	360 mmol
ECFV Contraction	10 L	24 mM	240 mmol
ECFV Restored	15 L	16 mM	240 mmol

1. *DECREASED NH_3 AVAILABILITY TO THE COLLECTING DUCT*. These patients have a low urine pH (<5.3).

2. *DECREASED COLLECTING DUCT H^+ SECRETION*. These patients have a urine pH of 5.8 or greater.

Decreased NH_3 Availability. There are 2 major categories for reduced NH_3 availability—cortical and medullary. In the former, hyperkalemia or a very low GFR result in reduced ammoniagenesis, whereas medullary diseases such as pyelonephritis and analgesic nephrotoxicity are common in the latter group. The distinctive features are metabolic acidosis and a normal plasma AG, low NH_4^+ excretion rates, and, unique to these patients, a very low urine pH (arbitrarily set at <5.3). This subgroup is also called hyperkalemic or Type IV RTA (see Tables 3–18 and 3–19, page 91).

Since their collecting duct H^+ secretory mechanism may be normal, these patients can have a very low urine pH. Nevertheless, they have little NH_3 available to bind the free H^+, and this reduces the overall capacity to excrete NH_4^+. This normal H^+ secretory mechanism may be indicated by a high urine P_{CO_2} (page 55) with alkali loading.

Impaired Collecting Duct H^+ Secretion. Collecting duct H^+ secretory defects can be considered in 4 subgroups (Table 3–19, page 91):

1. Problems with the H^+ ATPase pump;
2. Problems with voltage augmentation of this pump;
3. Low NH_3 availability to neutralize luminal H^+ (a high luminal $[H^+]$ lessens H^+ pumping);
4. An abnormal backleak of H^+ from the lumen into the cell.

Examples of each subgroup are shown in Table 3–19, page 91.

The above 4 subgroups have in common the low rate of NH_4^+ excretion. All but the third group above should have a high urine pH. However, pump failure (Group 1) should have a low urine P_{CO_2} in alkaline urine (<50 mm Hg; the others should be >70 mm Hg); hence, pump failure can also be distinguished. The low-voltage stimulation group tends to have reduced collecting duct Na reabsorption and, hence, to have a problem that interferes with K excretion (a low TTKG; see Fig. 9–5, page 223); these patients therefore have *hyperkalemia*, which lowers NH_4^+ production and excretion.

Diseases involving the renal medulla can destroy collecting duct cells, and thus these patients have low H^+ pump

TABLE 3–18. Comparison of Current and Proposed Classifications of RTA

Lesion: Reduced *Recycling* of Filtered HCO_3^-
CURRENT: Type II RTA, proximal RTA
PROPOSED: Proximal H^+ secretory defect
Lesion: Reduced Renal *New* HCO_3^- Generation
CURRENT: Type IV RTA
PROPOSED: Decreased NH_4^+ excretion due to hyperkalemia

CURRENT: Type I RTA, distal RTA, classic RTA
PROPOSED: Decreased NH_4 excretion due to low net distal H^+ secretion

TABLE 3–19. Steps Involved in Collecting Duct Net H^+ Secretion and Their Association With the Clinical Causes of RTA

Components for NH_4^+ Excretion	Causes of Lesions	Diagnostic Features
H^+ pump (ATPase)	Mineralocorticoid deficiency Medullary damage or infiltration ? Genetic	Low urine P_{CO_2} High urine pH
Voltage augmentation of H^+ secretion	Distal Na delivery low (i.e., ECF volume contraction, congestive heart failure, cirrhosis) Inhibitors of Na reabsorption (e.g., amiloride, lithium) Lack of stimulators (e.g., aldosterone deficiency or blockade) Excess Cl permeability	Low TTKG Hyperkalemia
Lowering luminal $[H^+]$ by NH_3 availability	Low NH_4^+ production Hyperkalemia, low GFR Low medullary NH_3 Tubulointerstitial disease	Low urine pH
Absence of backleak of H^+	Increased H^+ permeability (e.g., amphotericin B, ? toluene)	High urine pH High urine P_{CO_2}
Mechanism unclear	Dysproteinemias, outdated tetracycline	—

activity. This lesion also lowers NH_3 availability in the medulla. Hence, the urine NH_4^+ is low, but the urine pH can be <5.3 if the NH_3 defect predominates, >6 if the H^+ secretory defect predominates, or between 5 and 6, with neither predominating. In fact, a urine pH of 5.5 would suggest a combined NH_3 and H^+ defect.

Importance of the Plasma [K] in the Diagnosis of Distal RTA

Hypokalemia. In patients with distal RTA due to low collecting duct H^+ secretion rather than to low NH_3 availability, *hypokalemia* is often present. This hypokalemia is due to a higher than expected rate of K excretion and is associated with a urine pH that is >5.8.

Hyperkalemia: *Hyperkalemia* may be associated with metabolic acidosis and a low rate of NH_4^+ excretion; this condition has also been referred to as Type IV distal RTA (Table 3–18, page 91). The basis of the reduced NH_4^+ excretion is inhibition of NH_4^+ production by hyperkalemia. There are many reasons for the coexistence of hyperkalemia and metabolic acidosis, which can be divided into 3 subgroups (Table 3–20, page 93). Each circumstance can be identified by specific diagnostic features (Table 3–21, page 93). The first subgroup has insufficient distal Na resorption, owing to low Na delivery to the "cortical distal nephron" for aldosterone to act on; therefore, there is less K^+ and H^+ secretion here. A second subgroup has low aldosterone bioactivity; these patients have a low urine pH (Fig. 3–3, page 53). The low rate of NH_4^+ excretion disappears when hyperkalemia is treated. A third subgroup category has normal collecting duct H^+ secretion and aldosterone levels; however, they have another factor missing that makes them unable to generate a transtubular voltage. The most common of these defects is the failure to reabsorb filtered Na at the collecting duct, owing to a failure to respond to aldosterone or to the presence of drugs that block aldosterone action (e.g., amiloride). Rarely, a Cl shunt type lesion might be present.

Hyperkalemia and metabolic acidosis can also be present in renal failure.

OVERALL SUMMARY

Whereas metabolic acidosis may be the result of a large number of diverse disorders, these can be sorted out easily

TABLE 3–20. Impaired Voltage Augmentation of Collecting Duct H^+ and K^+ Secretion*

Low Distal Na^+ Delivery
- ECF volume contraction
- Hypoalbuminemia, cirrhosis, nephrotic syndrome

Low Aldosterone Bioactivity
- Prereceptor defects (e.g., adrenal gland or low renin)
- Receptor blockade (e.g., spironolactone)
- Postreceptor defects (e.g., amiloride, Cl shunt)

Decreased End Organ Function
- Interstitial nephritis
- Obstructive nephropathy
- Cyclosporin A toxicity

*All of these patients will have hyperkalemia associated with their metabolic acidosis.

TABLE 3–21. Hyperkalemia, Metabolic Acidosis, and Low Urine NH_4^+

Cause	Features	Basis
Decreased Na Reabsorption		
Low distal delivery	Urine [Na] <20 mM	ECFV depletion Severe hypoalbuminemia Congestive heart failure
Decreased mineralocorticoid	Urine pH <5.3 due to hyperkalemia, causing low NH_3 synthesis	Hyporeninemia Converting enzyme inhibitor Aldosterone antagonists Adrenal gland problem
Renal Failure	Low GFR	Low NH_4^+ production
Decreased Distal H^+ Secretion	Urine pH >5.7	Interstitial disease Drugs: amiloride, ? cyclosporin A
Chloride Shunt	ECF volume expansion Low urine [K] Urine pH <5.3 due to hyperkalemia	Dissipation of lumen negative transepithelial potential difference

with an organized approach, as outlined in the flow chart 3–D, page 95. Metabolic acidosis associated with increased H^+ production and with renal failure is identified by the increase in plasma AG. The subgroup due to methanol or ethylene glycol intoxication also has a wide osmolal gap. The flow charts detailing the group of patients with a wide plasma AG are on page 61.

The critical feature in sorting out the basis of metabolic acidosis with a normal plasma AG is whether or not the urine NH_4^+ excretion is appropriate. Urine $[NH_4^+]$ is reflected by the apparent urine net charge, i.e., $([Na^+] + [K^+]) - [Cl^-]$, when the urine pH is <6.1. When the urine apparent net charge is NEGATIVE (i.e., $[Cl^-] >> ([Na^+] + [K^+])$), this indicates appropriate urine $[NH_4^+]$, and the diagnosis is $NaHCO_3$ loss, either via the GI tract or some other site (proximal RTA, acetazolamide, NH_4Cl administration). If the urine apparent net charge is POSITIVE (i.e., $[Na^+] + [K^+] >> [Cl^-]$), the urine $[NH_4^+]$ is low, indicating the presence of RTA or diabetic ketoacidosis with marked β-hydroxybutyrate excretion along with NH_4^+, which will be revealed by the urine osmolal gap. The detailed flow chart for the diagnosis of RTA is on page 63.

Suggested Reading

General

Garella S, Dana CL, and Chazan JA: Severity of metabolic acidosis as a determinant of bicarbonate requirements. N Engl J Med 289:121, 1973.

Lennon EJ and Lemann J: Defense of hydrogen ion concentration in chronic metabolic acidosis. A new evaluation of an old approach. Ann Intern Med 65:265, 1966.

Narins RG and Emmett M: Simple and mixed acid-base disorders. A practical approach. Medicine 59:161, 1980.

Schwartz WB and Relman AS: Critique of the parameters used in evaluation of acid-base disorders. "Whole-blood buffer base" and "standard bicarbonate" compared with blood pH and plasma bicarbonate concentration. N Engl J Med 268:1382, 1963.

Anion Gap

Adrogue HJ, Brensilver J, and Madias NE: Changes in the plasma anion gap during chronic metabolic acid-base disturbances. Am J Physiol 235:291–297, 1978.

Emmett ME and Narins RG: Clinical use of the anion gap. Medicine (Baltimore) 56:38, 1977.

Gabow PA: Disorders associated with an altered anion gap. Kidney Int 27:472–483, 1984.

Halperin ML and Jungas RL: Metabolic production and renal disposal of hydrogen ions. Kidney Int 24:709–713, 1983.

Oh MS and Carroll HJ: The anion gap. N Engl J Med 297:814–817, 1977.

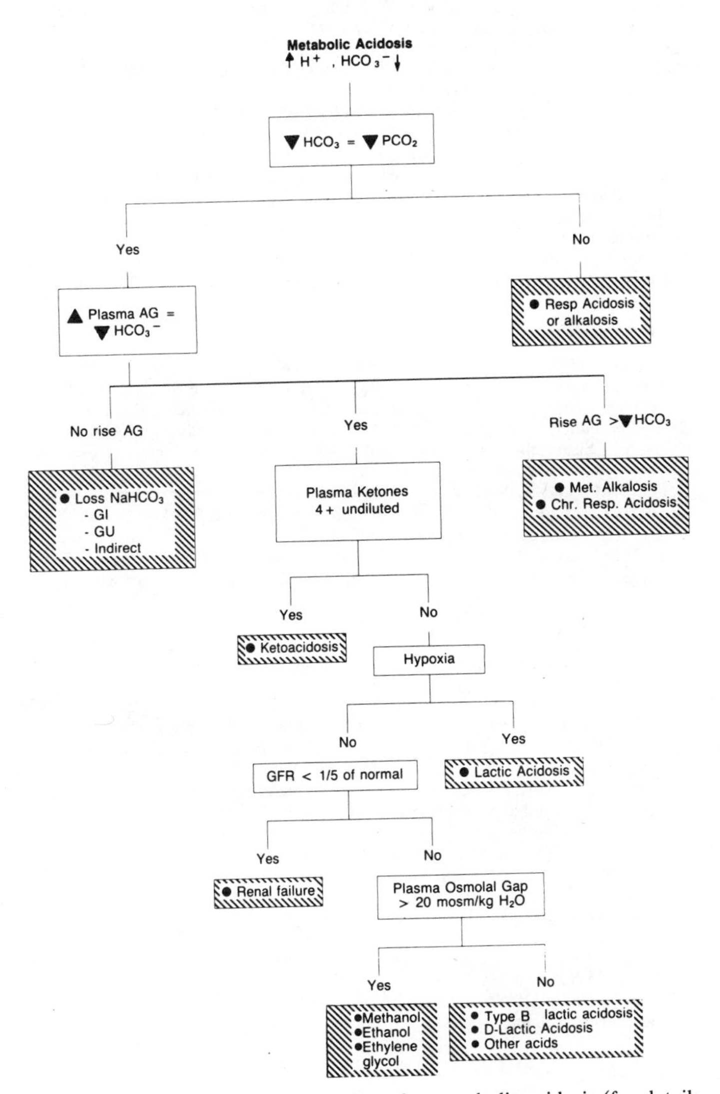

Flow Chart 3–D. Summary flow sheet for metabolic acidosis (for details see text). The solid triangles represent change in level, the open rectangles the parameters to evaluate, and the hatched boxes the final diagnoses.

Ketoacidosis

Adrogue HJ, Eknoyan G, and Suki WN: Diabetic ketoacidosis: Role of the kidney in the acid-base homeostasis re-evaluated. Kidney Int 25:591, 1984.
Cahill GF: Ketosis. Kidney Int 20:416, 1981.
Flatt JP: On the maximal possible rate of ketogenesis. Diabetes 21:50–53, 1972.
Halperin ML, Bear RA, Hannaford MC, et al: Selected aspects of the pathophysiology of metabolic acidosis in diabetes mellitus. Diabetes 30:781–787, 1981.
Halperin ML, Hammeke M, Josse RG, et al: Metabolic acidosis in the alcoholic: A pathophysiologic approach. Metabolism 32:308–315, 1983.
Hammeke M, Bear RA, Goldstein MB, et al: Hyperchloremic metabolic acidosis in diabetes mellitus. Diabetes 27:16–20, 1978.
Kreisberg RA: Diabetic ketoacidosis: New concepts and trends in pathogenesis and treatment. Ann Intern Med 88:681, 1978.
Marliss EB, Ohman JL, Aoki TT, et al: Altered redox state obscuring ketoacidosis in diabetic patients with lactic acidosis. N Engl J Med 283:978, 1970.

Lactic Acidosis

Cohen RD and Woods HF: Lactic acidosis revisited. Diabetes 32:181–191, 1983.
Fields AL, Wolman SL, and Halperin ML: Chronic lactic acidosis in a patient with cancer: Therapy and metabolic consequences. Cancer 47:2026–2029, 1981.
Fraley DS, Adler S, Bruns FJ, et al: Stimulation of lactate production by administration of bicarbonate in a patient with a solid neoplasm and lactic acidosis. New Engl J Med 303:1100–1102, 1980.
Halperin ML and Fields AL: Review: Lactic acidosis—Emphasis on the carbon precursors and buffering of the acid load. Am J Med Sci 289:154–159, 1985.
Madias NE: Lactic acidosis. Kidney Int 29:752–774, 1986.
Park R and Arieff AI: Lactic acidosis: Current concepts. Clin Endocrinol Metab 12:339–359, 1983.

Renal Tubular Acidosis

Defronzo RA: Hyperkalemia and hyporeninemic hypoaldosteronism. Kidney Int 17:118–134, 1980.
Goldstein MB, Bear RA, Richardson RMA, et al: The urine anion gap: A clinically useful index of ammonium excretion. Am J Med Sci 29:198–202, 1986.
Halperin ML, Goldstein ML, Richardson RMA, et al: Distal renal tubular acidosis syndromes: A pathophysiological approach. Am J Nephrol 5:1–8, 1985.
Halperin ML, Goldstein MB, Haig A, et al: Studies on the pathogenesis of Type I (distal) renal tubular acidosis as revealed by the urinary P_{CO_2} tensions. J Clin Invest, 53:669–677, 1974.
Rocher LL and Tannen RL: The clinical spectrum of renal tubular acidosis. Annu Rev Med 37:319–331, 1986.
Sebastian A, Schambelan M, Lindenfeld S, et al: Amelioration of metabolic acidosis with fludrocortisone therapy in hydroreninemic hypoaldosteronism. N Engl J Med 297:576, 1977.
Stinebaugh BJ, Schloeder FX, Goldstein MB, et al: Pathogenesis of distal renal tubular acidosis. Kidney Int 19:1–7, 1981.
Szylman P, Better OS, Chaimowitz D, et al: Role of hyperkalemia in the metabolic acidosis of isolated hypoaldosteronism. N Engl J Med 294:361–365, 1976.

Intoxications

Gonda A, Gault H, Churchill D, et al: Hemodialysis for methanol intoxication. Am J Med 64:749–757, 1978.
McCoy HG, Cipolle RJ, Ehlers SM, et al: Severe methanol poisoning. Am J Med, 67:804–807, 1979.
Parry ME and Wallach R: Ethylene glycol poisoning. Am J Med 57:143–150, 1974.
Peterson CD, Collins AJ, Bullock ML, et al: Ethylene glycol poisoning. New Engl J Med 304:21–23, 1981.

4
METABOLIC ALKALOSIS

THE BIG PICTURE

Metabolic alkalosis is present when there is a rise in the [HCO_3^-] together with a fall in the [H^+]. Having an increased [HCO_3^-] involves either a source of new HCO_3 or ECF volume contraction (with no change in HCO_3 content) and renal mechanisms to permit a sustained elevation in the plasma [HCO_3^-].

Clinically, there are 2 major subgroups: (1) THE SALINE-SENSITIVE GROUP—the common group that responds to Cl administration (KCl and NaCl); (2) THE SALINE-RESISTANT GROUP that does not respond to saline.

Saline-sensitive metabolic alkalosis is caused most often by vomiting or diuretics; this condition is characterized by a very low Cl excretion rate in the urine (when diuretics are not acting).

Patients may not admit that they induce vomiting or abuse diuretics. A high degree of suspicion is required.

Cl-resistant metabolic alkalosis is less common. It occurs with high aldosterone states, Mg depletion, Bartter's syndrome, and in renal failure with a Na-HCO_3 load. Therapy is usually more difficult with this type.

GENERATION OF NEW HCO_3^-

There are two organ systems capable of generating new HCO_3^-: the upper GI tract and the kidneys.

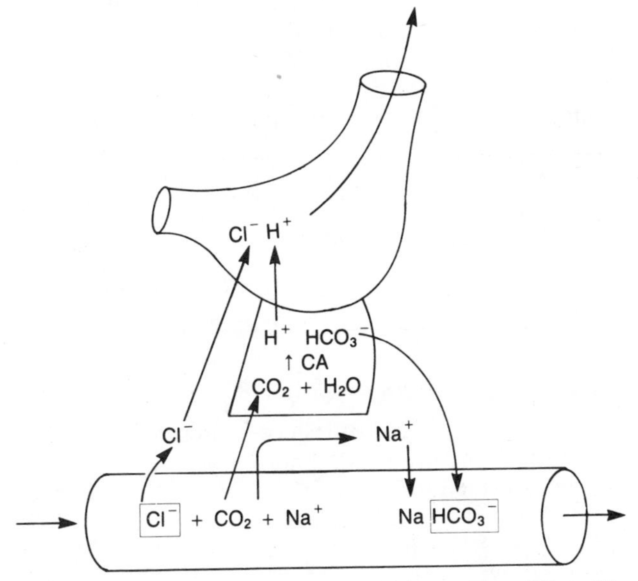

Figure 4–1A. New HCO_3^- formation with vomiting. When H^+ are secreted into the gastric lumen, HCO_3^- ions are added to the venous blood. If the gastric contents were removed, there would be a large positive HCO_3^- balance.

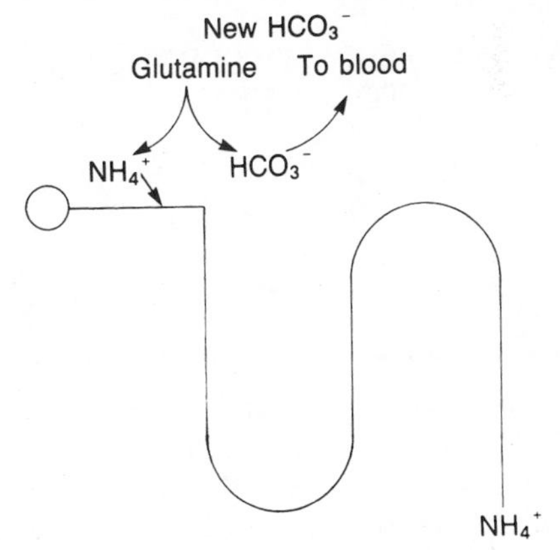

Figure 4–1B. Renal new HCO_3^- formation. Glutamine is metabolized in proximal tubular cells to produce NH_4^+ and HCO_3^- ions. For a net gain of new HCO_3^- in the body, these NH_4^+ ions must be excreted in the urine.

ILLUSTRATIVE CASE

A 26 year old dancer complains of weakness. She denies vomiting and the intake of drugs other than vitamins. Physical examination reveals a thin woman with a blood pressure of 95/60 mm Hg and a postural drop in BP of 15 mm Hg. The jugular venous pulsations cannot be seen. Laboratory results are listed below:

	Plasma	Spot Urine
Na	137	52
K	3.1	50
Cl	90	0
HCO_3	32	Not determined
pH	7.48	8.0

Questions (Answers on page 101)

1. What acid-base disturbance is present?
2. Why is the urine [Na^+] not lower, given the ECF volume contraction?
3. Why is she hypokalemic?
4. What is the underlying cause for the acid-base disturbance?

PATHOPHYSIOLOGIC CONSIDERATIONS

METABOLIC ALKALOSIS REQUIRES RENAL MECHANISMS TO SUSTAIN A HIGH ECF [HCO_3^-] AND EITHER:

1. NEW HCO_3 ADDITION (GASTRIC H^+ LOSS, RENAL NH_4^+ EXCRETION, INTAKE OF HCO_3^- OR CITRATE $^-$), OR
2. LOSS OF ECF NaCl PLUS WATER (KIDNEY, GI, OR SKIN)

Metabolic alkalosis is a disorder characterized by an increased ECF [HCO_3^-] and a reduced [H^+]. Two pathophysiologic processes must be explained:

1. Mechanisms whereby the ECF [HCO_3^-] is elevated;
2. A renal basis to maintain an elevated [HCO_3^-].

Answers

1. The patient has METABOLIC ALKALOSIS, increased plasma [HCO_3^-], and alkalemia.

2. For the urine [Na] to be low, the filtered Na must be reabsorbed. There are 3 ways in which Na can be reabsorbed: along with Cl, and in conjunction with K^+ or H^+ secretion into the lumen. In response to ECF volume contraction, the urine Cl is *close to* 0, indicating appreciable NaCl reabsorption. However, in this case, some Na was not reabsorbed; it was excreted along with a non-Cl^- anion (note the urine net charge of 102 mEq/L). Since the urine pH is very high, this indicates that the "nonreabsorbed anion" is HCO_3^-. To the degree that the filtered load of HCO_3 exceeds the tubular capacity to secrete H^+, the HCO_3^- is excreted, obligating the coincident excretion of K^+ or Na^+.

3. She is hypokalemic because she has ECF volume contraction, resulting in mineralocorticoid release. Mineralocorticoids promote K^+ secretion and Na^+ reabsorption in the cortical collecting tubule. Adequate Na delivery to this segment is ensured by the presence of the nonreabsorbable anion, HCO_3^-. The high urine [K] in the presence of hypokalemia confirms the action of mineralocorticoids.

4. The patient has metabolic alkalosis and coincident renal HCO_3^- excretion. If this represents a steady state (continued metabolic alkalosis despite bicarbonaturia) she must have an ongoing supply of HCO_3^- (you cannot lose HCO_3^- and maintain an elevated plasma level without an ongoing source). If the source were exogenous $NaHCO_3$ ingestion, there would be ECF volume expansion, which is not the case. Hence, there is endogenous generation of HCO_3^- that is not of renal origin (there is renal HCO_3 loss, not generation). Therefore, the patient must be vomiting. An alkaline urine pH with a very low urine [Cl] in a patient with metabolic alkalosis is pathognomonic of vomiting.

Elevation of ECF [HCO_3^-]

There are 2 ways to raise the ECF HCO_3^-/water ratio:

1. Increase the numerator of this ratio by adding new HCO_3^-;
2. Reduce the denominator of this ratio by contracting the ECF volume by NaCl removal.

New HCO_3^- Generation

The common ways in which new HCO_3^- generation occur are either via the GI tract or the kidneys (see page 99). Alternatively, a patient might be given either HCO_3^- itself or an organic anion that is metabolized to HCO_3^- (e.g., citrate or lactate).

Contraction of the ECF Volume

It is possible for the level of ECF [HCO_3^-] to rise without adding new HCO_3^- if the ECF volume is contracted by the removal of NaCl and water. In this situation, the [HCO_3^-] rises because the HCO_3^- is distributed in a smaller volume—i.e., the CONTENT is unchanged, but the CONCENTRATION is increased (see Answer to Question 8, page 105). Diuretic ingestion is commonly associated with metabolic alkalosis; these agents cause NaCl loss in the urine and ECF volume contraction (they may also cause new HCO_3 formation in the kidney by increasing NH_4^+ excretion by indirect mechanisms).

Questions (Answers on page 103)

5. If a patient has gastric drainage, can the quantity of new HCO_3^- formation be minimized?
6. Why does the plasma [HCO_3^-] rise in a patient with normovolemia who is undergoing plasmapheresis for rapidly progressive glomerulonephritis and renal failure?

Mechanisms for Renal HCO_3^- Retention

If a normal person were to ingest substantial amounts of $NaHCO_3$, the ECF [HCO_3^-] would temporarily increase by only a small amount before all of this extra HCO_3 would be

Answers

5. The patient with gastric drainage can lose large amounts of HCl, resulting in metabolic alkalosis. The most effective way to reduce the H^+ loss is to inhibit H^+ secretion by the gastric mucosa by administering H_2 receptor blockers. Although this markedly reduces the HCl loss, significant NaCl or $NaHCO_3$ can be lost, and this must be replaced to avoid ECF volume contraction and a change in plasma [HCO_3^-] on that basis.

The other ways in which one might reduce the acid-base impact of large HCl losses are either to keep the patient's ECF volume well expanded with NaCl so that the $NaHCO_3$ is excreted (if renal function is adequate), or if the renal function is poor, to administer H^+ in the form of HCl or NH_4Cl to titrate the excess HCO_3^-. Titration of HCO_3^- with H^+ is a CO_2-producing reaction. This might lead to CO_2 retention in patients with compromised ventilation.

6. The patient's ECF [HCO_3^-] is increasing with no evidence of ECF volume contraction, indicating an increased ECF HCO_3^- content. In the absence of endogenous HCO_3^- generation (no vomiting and renal failure, which prevent renal HCO_3^- generation), the extra HCO_3 must be due to exogenous bicarbonate, HCO_3^- mobilized from an ileus, or the metabolic production of HCO_3^-. There was no ileus and the patient was not receiving HCO_3^- per se. However, daily plasmapheresis with plasma replacement (and several blood transfusions) provided a large Na citrate load (the anticoagulant) that, when metabolized to CO_2, consumes H^+, yielding HCO_3^-. The renal failure prevented appreciable $NaHCO_3$ excretion.

excreted at a rapid rate (for mechanisms, see Chapter 1, pages 16 and 18). Since patients with metabolic alkalosis often have a plasma [HCO_3^-] in excess of 35 mM, they must have an alteration of the renal mechanisms to enable them to maintain this very elevated ECF [HCO_3^-]. There are 2 mechanisms that might permit such a high plasma [HCO_3^-] in the presence of normal kidney function:

1. Enhanced HCO_3^- reabsorption.
2. Decreased glomerular filtration rate (GFR).

It has generally been assumed that the renal basis for the elevated ECF [HCO_3^-] is increased reabsorption of filtered HCO_3^-, with the alleged stimuli being ECF volume contraction, Cl depletion, and hypokalemia. However, it has been difficult to substantiate an appreciable increase in ABSOLUTE HCO_3^- reabsorption in animal models. Therefore, an additional component to the renal preservation of the elevated [HCO_3^-] must be a reduction in the filtered load of HCO_3^-. There is increasing evidence that there is a nearly proportionate reduction in the GFR to keep the filtered load of HCO_3^- near normal (i.e., when the plasma [HCO_3] doubles, the GFR almost halves). This is an important component of the renal contribution to maintain the elevated plasma [HCO_3] in metabolic alkalosis. The basis of the decreased GFR in many cases is ECF volume contraction. However, in other cases, such as the metabolic alkalosis consequent to mineralocorticoid excess, there is no ECF volume contraction. Nevertheless, decreased GFR has been demonstrated in animal models of this disorder, and this has been attributed to the associated hypokalemia. In yet other cases, the drop in GFR may be due to Cl depletion.

Thus, in the approach to the correction of metabolic alkalosis, one has to correct ECF volume depletion and replace the K and Cl deficits. In addition, one should deal with the original basis of the metabolic alkalosis.

Questions (Answers on page 105)

7. Why is the urine [Na] not a reliable indicator of ECF volume contraction in patients who vomit?

8. What is the impact on the ECF HCO_3^- CONTENT and CONCENTRATION of the loss of 200 mmol of Na and Cl due to diuretic action? (For simplicity, assume no change in [Na^+], which is 140 mM.)

Answers

7. With vomiting, there is a net loss of ECF Cl^-, which is replaced by HCO_3^- (Fig. 4–1, page 99). Given the same ECF volume and GFR, the filtered load of HCO_3^- rises, but proximal H^+ secretion does not rise (no depletion of ECF volume, K, or Cl, but alkalemia is present). Hence this increased filtered load of HCO_3^- results in $NaHCO_3^-$ excretion, which leads to a small degree of ECF volume contraction. The urine contains Na despite the ECF volume depletion because it is accompanying HCO_3^-, which cannot be reabsorbed. With the next episode of vomiting, the same sequence of events occurs—the ECF $[HCO_3^-]$ rises further, and some bicarbonaturia occurs along with some Na^+ and more K^+ loss (the low ECF volume leads to aldosterone release). This time the ECF $[HCO_3^-]$ is slightly higher than previously. The degree of ECF volume depletion depends on the Na intake between episodes of vomiting and the magnitude of the Na^+ losses (both GI and renal). With each episode of vomiting, there is transient bicarbonaturia as the "renal threshold" for HCO_3^- reabsorption is exceeded by the gastric HCO_3^- generation, but the ECF $[HCO_3^-]$ progressively rises, since the bicarbonaturia is not sufficient to excrete all the HCO_3^- generated. The progressive ECF volume depletion limits the amount of HCO_3^- excretion in 2 ways: first, owing to decreased GFR, the amount of HCO_3^- filtered per unit of time (the filtered load) is decreased, and secondly, by stimulating mineralocorticoid release, collecting duct H^+ secretion is facilitated, resulting in increased HCO_3^- reabsorption and increased K^+ excretion. Therefore, the patient with recurrent vomiting will have metabolic alkalosis, hypokalemia, and ECF volume contraction.

The urine composition varies, depending on when the patient is evaluated. After the patient vomits, the urine contains Na and K HCO_3 but no chloride (owing to ECF volume contraction and Cl depletion, all the Cl is reabsorbed), and the urine pH is alkaline. Between episodes, the urine contains little Na^+ or Cl^+ (owing to ECF volume contraction) and some K^+ (excreted with $SO_4^=$ and $HPO_4^=$). The urine pH is low, due to mineralocorticoid enhancement of H^+ secretion.

8. If we assume the patient weighs 70 kg with an ECF $[Na^+]$ of 140 mM, the total body Na is 14 L $\times$ 140 mM $=$ 1960 mmol. If 200 mmol Na are lost, the total body Na is now 1760 mmol, and if the ECF $[Na^+]$ remains 140 mM, the ECF volume is now 12.6 L. The HCO_3^- content was 25 mM $\times$ 14 L $=$ 350 mmol, which is unchanged; thus, the patient's current ECF $[HCO_3^-]$ $=$ 350/12.6 $=$ 28 mM.

Therefore, you can appreciate that with the loss of NaCl the ECF $[HCO_3^-]$ can rise WITH NO CHANGE IN ECF HCO_3^- CONTENT, owing to a decrease in ECF volume. This process contributes to the metabolic alkalosis due to diuretic use.

CLINICAL SETTINGS

METABOLIC ALKALOSIS OCCURS MOST COMMONLY WITH:
1. LOSS OF GASTRIC CONTENTS
2. DIURETIC USE
3. MINERALOCORTICOID EXCESS

Metabolic alkalosis is a common acid-base disorder occurring in the above 3 major clinical settings. In view of these 3 causes, one might think that the diagnosis of metabolic alkalosis would be straightforward, and often it is. However, although the first 2 categories account for the vast majority of cases, a number of patients in each, because of personality disorders, are not willing to admit to their activities and deny the self-induction of vomiting or the abuse of diuretics. Therefore, the history cannot be relied upon entirely, and one must often resort to a series of laboratory tests and a degree of subterfuge to establish the diagnosis.

DIAGNOSTIC CATEGORIES

From the diagnostic and therapeutic point of view, it is useful to divide causes of metabolic alkalosis into 2 groups (see Tables 4–1 and 4–2, page 107).

GROUPS OF METABOLIC ALKALOSIS
1. CHLORIDE-RESPONSIVE
2. CHLORIDE-RESISTANT

Chloride-Responsive Metabolic Alkalosis

- ECF volume contraction
- Urine [Cl] <20 mM, unless on diuretics

These patients are called "Cl-responsive" because their metabolic alkalosis can be corrected by Cl^- administration. The major diagnostic features of this group are that their urine is virtually free of Cl^- (<20 mM) and that they usually

TABLE 4–1. Causes of Chloride-Responsive Metabolic Alkalosis

- GASTRIC FLUID LOSS
- DIURETIC USE
- NONREABSORBABLE ANION DELIVERY
- POSTHYPERCAPNIA
- STOOL Cl^- LOSS
  - VILLOUS ADENOMA
  - CONGENITAL CHLORIDORRHEA

TABLE 4–2. Causes of Chloride-Resistant Metabolic Alkalosis

- PRIMARY HYPERALDOSTERONISM
- SECONDARY HYPERALDOSTERONISM
  - MALIGNANT HYPERTENSION
  - RENAL ARTERY STENOSIS
  - RENIN-SECRETING TUMOR
  - MAGNESIUM DEPLETION
- BARTTER'S SYNDROME
- GLUCOCORTICOID HORMONE EXCESS

PATHOGENESIS OF DIURETIC-INDUCED METABOLIC ALKALOSIS

In diuretic-induced metabolic alkalosis, 2 processes play a role; the first is contraction of the ECF volume. Not only does this increase the ECF [HCO_3^-] (see Answer 8, page 105), but owing to the associated decrease in GFR, it reduces the filtered load of HCO_3^-, so that bicarbonaturia does not result in return of the [HCO_3^-] to normal.

The second process is increased renal NEW HCO_3^- generation. The associated hypokalemia in diuretic use stimulates ammoniagenesis, increasing HCO_3^- formation (see Chapter 1, page 23). The ECF volume depletion and subsequent stimulation of mineralocorticoid release result in a significant increase in renal H^+ secretion (preserving the new HCO_3^- that had been generated).

As with metabolic alkalosis due to vomiting, the composition of the urine varies in the patient on diuretics, depending on the time of observation. If the diuretic ingestion has been recent (within 6 hr), the urine contains substantial amounts (>20 mM) of Na and Cl (and K if the diuretic was not a K-sparing variety). On the other hand, if the diuretic ingestion was more remote, there will be very little Cl or Na in the urine. Fluctuating urine Na and Cl concentrations (from <10 mM to >30 mM) in a patient with ECF volume contraction is a clue that there is intermittent diuretic ingestion (see Table 4–4, page 115).

have ECF volume contraction. However, if the patient has just taken a diuretic, the urine contains significant amounts of Cl despite the fact that he or she belongs to this category (see page 107). In some experimental models of metabolic alkalosis, the GFR may be low with no obvious reduction in ECF volume; this seems to be an intrarenal effect of Cl depletion.

The commonest causes of metabolic alkalosis in general, and of the saline-sensitive variety in particular, are either diuretic use or loss of gastric secretions. The additional causes are:

Nonreabsorbable Anions. If a patient has ECF volume contraction and takes a Na salt with an anion that cannot be reabsorbed by the kidney (e.g., Na carbenicillinate), the patient may develop metabolic alkalosis (see page 109 for pathophysiology).

Stool Cl Loss. The loss of Cl-rich fluid in the stool (e.g., congenital chloridorrhea, a rare disorder) creates a situation identical to the diuretic-induced metabolic alkalosis, except that the urine always has a low [Na] and [Cl].

Posthypercapnia. In the course of chronic hypercapnia, there was Cl loss in the urine when the patient developed the increased [HCO_3^-], i.e., enhanced renal NH_4Cl excretion. If the patient has a contracted ECF volume when the hypercapnia resolves, there will be a stimulus for Na reabsorption and H^+ secretion, reabsorbing luminal HCO_3^-. The increased plasma [HCO_3^-] is thus maintained until the ECF volume is reexpanded with NaCl administration.

Cl-Resistant Metabolic Alkalosis

- Urine Cl >20 mM
- Not ECF volume contracted in most cases
- Usually hypertensive

Metabolic alkalosis of this variety is much less common. It is not corrected by NaCl or KCl administration. The urine contains appreciable amounts of Cl if a normal diet is eaten. These patients usually do not have ECF volume contraction and they tend to be hypertensive. An exception to this rule are those patients with either Bartter's syndrome or Mg

PATHOGENESIS OF NONREABSORBABLE ANION-RELATED METABOLIC ALKALOSIS

The patient who develops metabolic alkalosis in association with the ingestion of a nonreabsorbable anion has ECF volume contraction for some unrelated reason. Thus, there is a stimulus to Na reabsorption (ECF volume contraction), but the Na cannot be reabsorbed in the proximal tubule (anion nonreabsorbable). Therefore, increased amounts of Na are delivered to the distal nephron, where again the anion is not reabsorbed. Therefore, the Na is reabsorbed in conjunction with H^+ and K^+ secretion, resulting in hypokalemia and an increased plasma [HCO_3^-] (a result of renal HCO_3^- generation with perhaps a component due to ECF volume contraction).

The urine provides the clues to the diagnosis:

1. URINE [Cl] LOW (<20 mM).
2. URINE [Na] VARIABLE (if large recent Na load with a nonreabsorbable anion was administered, the urine [Na] will exceed 30 mM; if Na intake is small, the urine [Na] may be <10 mM).
3. Substantial unmeasured anion concentration in the urine ([Na + [K] >> [Cl]). If the urine pH is alkaline, the anion is HCO_3^-; if acid, the patient has a nonreabsorbable anion-related metabolic alkalosis.

PATHOPHYSIOLOGY OF CHLORIDE-RESISTANT METABOLIC ALKALOSIS

In all cases of Cl-resistant metabolic alkalosis, there is delivery of Na and Cl to the cortical collecting duct in the presence of mineralocorticoid activity, which results in the reabsorption of Na and the excretion of Cl, H^+, and K^+. As a result of the K excretion, hypokalemia develops, which results in stimulation of ammoniagenesis and new HCO_3^- generation. Thereafter, increased H^+ excretion in the form of NH_4^+ results in the preservation of metabolic alkalosis. In addition, there may also be a decrease in the GFR (secondary to the hypokalemia), resulting in a decrease in the filtered load of HCO_3^-, which further promotes the maintenance of the metabolic alkalosis. The basis of the mineralocorticoid activity varies, depending on the diagnosis, as outlined below:

Diagnosis	Basis of Mineralocorticoid Activity
Bartter's Syndrome	ECF volume contraction (Na wasting)
1° Aldosteronism	Autonomous secretion (adenoma hyperplasia)
Glucocorticoid Excess	Autonomous ACTH secretion
Renal Artery Stenosis	Secondary to hyperreninemia
Renin-Secreting Tumor	Secondary to hyperreninemia
Drug-Induced	Exogenous mineralocorticoid activity
Magnesium Depletion	Mechanism unclear

depletion: they may have ECF volume contraction and normotension.

Hyperaldosteronism. The majority of patients with Cl-resistant metabolic alkalosis have excessive mineralocorticoid activity of either exogenous or endogenous origin (Table 4–2, page 107). The associated hypokalemia is probably of major importance in both the generation and the maintenance of the metabolic alkalosis, both by enhancing ammoniagenesis, which enables renal new HCO_3^- formation, and by reducing the GFR, enabling maintenance of the elevated blood $[HCO_3]$.

Bartter's Syndrome. Patients with this unusual disorder are not hypertensive and have ECF volume depletion, renal NaCl wasting, metabolic alkalosis, and hypokalemia. The exact basis of the Na and Cl wasting remains controversial; they either have more Na and Cl delivered to the mineralocorticoid-sensitive area of the nephron than it can reabsorb, or they have impaired Na and/or Cl reabsorption at this nephron site.

Magnesium Depletion. Patients with Mg depletion may have metabolic alkalosis and hypokalemia resembling a high mineralocorticoid state. The hypomagnesemia can be confirmed by plasma analysis. The usual clinical setting for this deficiency is in a patient with malabsorption or diarrhea or in the administration of drugs acting on the loop of Henle, such as cisplatin, diuretics, or aminoglycosides. These patients must be distinguished from those with primary hyperaldosteronism who may also have Mg deficiency.

Alkali Loading. Under usual circumstances, $NaHCO_3$ loading leads to only a mild elevation in the plasma $[HCO_3^-]$, as the bulk of this HCO_3^- is excreted. However, in the presence of Na depletion or in renal failure, significant elevations of plasma $[HCO_3^-]$ occur with $NaHCO_3$ administration.

Milk-Alkali Syndrome. The milk-alkali syndrome (metabolic alkalosis, hypercalcemia, hypocalciuria, and renal insufficiency) is due to the ingestion of large amounts of milk and absorbable antacids ($CaCO_3$). It has been mainly of historical interest; however, with the current emphasis on osteoporosis prevention with $CaCO_3$ as a major source of Ca supplementation, we may see a reappearance of this disorder. The pathogenesis is detailed on page 111.

Nonreabsorbable Alkali Ingestion With Ion Exchange Resins. Ion exchange resins are generally used in patients

PATHOGENESIS OF METABOLIC ALKALOSIS

Milk-Alkali Syndrome. The basis of this disorder is the ingestion and absorption of large amounts of calcium and HCO_3^-. The patients then excrete large amounts of calcium and HCO_3^- in their urine leading to nephrocalcinosis (calcium oxalate is less soluble in alkaline urine) and renal failure. As renal function diminishes, the ability to excrete the large calcium load is reduced. These patients actually have hypocalciuria, and they develop hypercalcemia with soft tissue calcification. The hypercalcemia may cause vomiting, leading to metabolic alkalosis; it also stimulates proximal HCO_3^- recycling; finally, there is impaired renal HCO_3^- excretion due to the decreased renal function. The metabolic alkalosis and hypercalcemia resolve when ingestion of the calcium and the alkali ceases.

Nonabsorbable Alkali and Ion Exchange Resin Ingestion. The proposed mechanisms of the metabolic alkalosis are:

1. Intake of Ca, Mg, or Al salts of organic anions.
2. Intake of resin that is *not* in H^+ form (i.e., with Na or K).
3. Metabolism of organic anion to HCO_3^-.

4a. Normally, Ca, Mg, or Al are excreted in the feces as carbonate salts consuming HCO_3, and acid-base balance results.

4b. If Ca, Mg, or Al form a complex with the resin in the GI tract, there is no GI HCO_3 loss.

4c. If option (4b) rather than (4a) occurs and the patients have a very low GFR, they cannot excrete the HCO_3^- and become progressively alkalemic.

IMPACT OF METABOLIC ALKALOSIS ON PATIENTS WITH CO_2 RETENTION

Patients with chronic lung disease often take diuretics to cope with their Na^+ retention and may develop metabolic alkalosis. The mixed acid-base disturbance may return their plasma $[H^+]$ to the normal range; this removes the acidemic drive to ventilate and may worsen the clinical condition. It is important to note that these patients need not be alkalemic in order to experience the adverse effects of metabolic alkalosis on chronic respiratory acidosis. The data below represent 11 episodes of metabolic alkalosis in 8 patients with chronic respiratory acidosis prior to and following the correction of the metabolic alkalosis. No change in the status of the respiratory function tests occurred.

	$[H^+]$ (nM)	HCO_3^- (mM)	P_{CO_2} (mm Hg)	P_{O_2} (mm Hg)
Before Correction	40	37	61	52
After Correction	42	28	48	69

with renal insufficiency; when combined with nonreabsorbable alkali (aluminum or magnesium salts) their use has resulted in metabolic alkalosis that resolves when either agent is discontinued. The pathogenesis is described on page 111.

EFFECT OF METABOLIC ALKALOSIS ON VENTILATION

The $[H^+]$ is a major determinant of ventilation, and therefore one would expect metabolic alkalosis to have a depressant effect on ventilation. In fact, there is a linear relationship between the increasing plasma $[HCO_3^-]$ and progressive increase in P_{CO_2}, with a slope of approximately 0.7, indicating that for every mM increase in plasma $[HCO_3^-]$, you should expect to see a 0.7 mm Hg increase in P_{CO_2}. Thus, when patients present with CO_2 retention and metabolic alkalosis, the metabolic alkalosis should be corrected prior to attributing the CO_2 retention to lung disease.

As hypoventilation develops, it is accompanied by hypoxia, which offsets the degree of respiratory suppression achieved (more severe respiratory suppression is observed in patients receiving O_2 supplementation when hypoxia is prevented). The reduction of tissue O_2 delivery in metabolic alkalosis is further aggravated by the fact that the O_2 hemoglobin dissociation curve is shifted to the left by alkalemia, increasing the affinity of hemoglobin for O_2.

Patients with chronic respiratory acidosis may be adversely affected by a coexistent metabolic alkalosis (a common phenomenon, as these patients often use diuretics), since less severe acidemia further suppresses their ventilation (see page 111 for further discussion).

THE DIAGNOSTIC APPROACH TO THE PATIENT WITH METABOLIC ALKALOSIS (see Flow Chart 4–A, page 113)

The first step in the resolution of the basis of metabolic alkalosis is to rule out the presence of significant renal function impairment (GFR <25%). If renal failure exists, the specific cause of the metabolic alkalosis should be evident

The clinical condition (mental function and sense of well-being) was improved by both the fall in P_{CO_2} and the increase in P_{O_2}. Note the substantial improvement in P_{CO_2} and P_{O_2} despite the modest change in [H+].

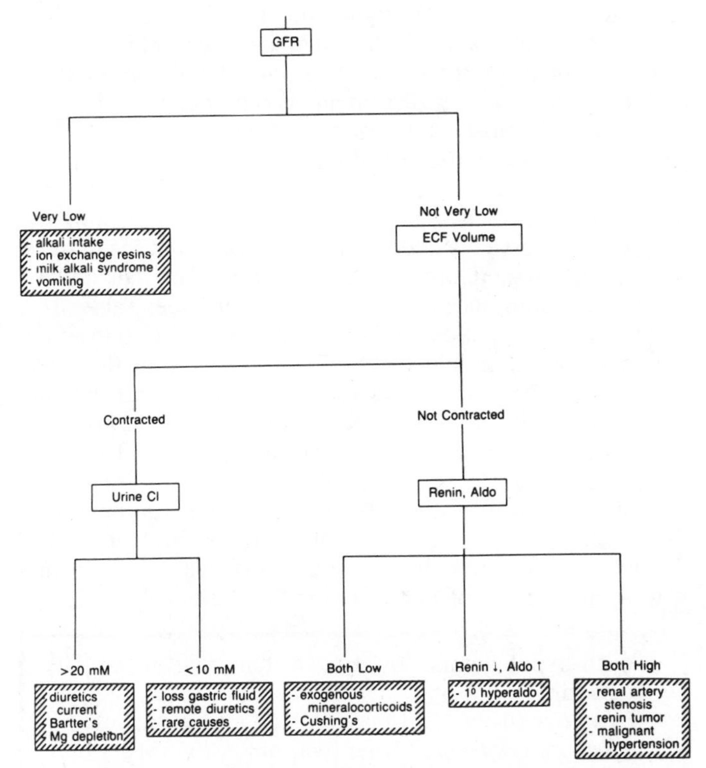

Flow Chart 4–A. Diagnostic flow chart for metabolic alkalosis. The information to be interpreted to make a final diagnosis is shown in the open boxes; the final diagnoses are shown in the shaded boxes (see text for details).

from the history.

If a very low GFR plus alkali input are not the cause of metabolic alkalosis, the ECF volume status is the next critical parameter to assess. The majority of patients have a contracted ECF volume and their urine [Cl^-] is very low (<20 mM). The differential diagnosis in this ECF volume-contracted group is indicated in Flow Chart 4–A on page 113, and the criteria for identifying the possible causes if the history is not available are in Table 4–3, page 115.

There are 2 groups of patients with metabolic alkalosis and ECF volume contraction in whom the urine [Cl^-] is >20 mM; the most likely cause is diuretic use (abuse). If these patients deny the use of diuretics, they can usually be identified by measuring the electrolytes on serial random urine samples. Some will be Na- and Cl-free, indicating the ability to conserve Na and Cl (more remote from the diuretic consumption), and others will have abundant Na and Cl (following more recent diuretic ingestion; see Table 4–4, page 115). The patient who has ECF volume contraction and wasting Na should have excessive K loss in the urine (owing to mineralocorticoid secretion due to ECF volume contraction); if the urine [K^+] is not high, suspect the ingestion of a diuretic with "K-sparing" qualities. The other clues suggesting the diagnosis of diuretic abuse are: access to diuretics (paramedical personnel), "plump" patients with disturbed "body image," elevated serum uric acid, and fluctuating body weight. The diagnosis can be confirmed with a direct assay of the urine for the diuretic.

- With hypokalemia, low ECF volume and urine [Cl] >20 mM, suspect:
  Diuretics (urine [Na] may be <20 mM at times)
  Bartter's syndrome (urine [Na] not <20 mM)

The other group of patients with metabolic alkalosis, ECF volume depletion, and inappropriately high urine Na and Cl excretions are patients with Bartter's syndrome. Serial observations of these patients reveal persistent Na and Cl wasting and a relatively constant body weight, both of which contrast with a diagnosis of diuretic abuse.

TABLE 4–3. Clues for Identifying the Possible Causes of Cl-Responsive Metabolic Alkalosis

Cause	Diagnostic Features
Vomiting (recent)	• UNMEASURED ANION IN URINE (HCO_3) • URINE pH >7, [Cl] LOW, [Na] HIGHER • DISTURBED BODY IMAGE
Remote Diuretic ***Posthypercapnia*** ***Cl Loss In Diarrhea***	• URINE [Na] AND [Cl] LOW • NO UNMEASURED ANION IN URINE • URINE pH <6
Nonreabsorbable Anion	• URINE [Cl] LOW, BUT [Na] HIGHER • UNMEASURED ANION IN THE URINE • URINE pH <7

TABLE 4–4. Pattern of Serial Urine Electrolyte Determinations in a Patient With ECF Volume Contraction, Indicating Diuretic Abuse

	A*	B†	C‡
[Na] (mM)	30	10	42
[K] (mM)	43	13	59
[Cl] (mM)	85	3	108

*Indicates modest Na and Cl wasting during ECF volume contraction, suggesting diuretic action.

†Na and Cl conservation, no diuretic effect remaining.

‡Na and Cl wasting now evident again, despite ECF volume depletion in a patient previously capable of conserving Na and Cl, indicating recent diuretic ingestion. The kaliuresis despite hypokalemia confirms the clinical impression of ECF volume depletion (evidence of mineralocorticoid secretion).

If one uses the jugular venous pressure as the primary indicator of ECF volume status (the adequacy of the circulating volume is actually being assessed), one can be misled by conditions that raise the central venous pressure in the presence of low left ventricular output (Table 4–5, page 117).

In patients without ECF volume contraction (they are also usually hypertensive and have Cl in their urine), the major problem is to establish the basis of the increased mineralocorticoid activity (whether primary hyperaldosteronism or a secondary phenomenon). This determination is easily made by coincident plasma renin and aldosterone levels and by measuring the serum magnesium level as indicated in Flow Chart 4–A, page 113.

TREATMENT OF METABOLIC ALKALOSIS

Cl-Responsive Group

- Replace Na, K, and Cl deficits
- Identify basis of deficits

Patients with ECF volume contraction and Cl and K depletion respond rapidly to Na, K, and Cl replacement. When their ECF volume is restored, the $[HCO_3^-]$ falls, owing to dilution; eventually there is bicarbonaturia, resulting in the excretion of any excess HCO_3^-. Generally, the metabolic alkalosis responds to NaCl therapy, even if the K deficits are not restored. However, there are a few patients who have been described with very large K deficits ($[K^+]$ <2.4 mM) who were resistant to NaCl administration until the K deficits were replenished.

In patients in whom the metabolic alkalosis is due to a reduction in "effective circulating volume" with ECF volume expansion (e.g., congestive heart failure on diuretics), the metabolic alkalosis generally improves with KCl therapy, and additional Na should *not* be given (as K enters the ICF, Na leaves, adding some Na to the ECF).

If the patient has been abusing diuretics or inducing vomiting, therapy of the underlying psychopathology proves to be the greater problem.

TABLE 4–5. Conditions in Which Jugular Venous Pressure (JVP) Does Not Reflect Circulating Volume

Pulmonary Hypertension
Chronic emphysema
Pulmonary emboli
Idiopathic pulmonary hypertension
Impaired Right Ventricular Emptying
Pulmonary valve stenosis or insufficiency
Cardiomyopathy
Right ventricular infarction
Tricuspid valve insufficiency
Impaired Right Ventricular Filling
Tricuspid valve stenosis
Superior vena cava syndrome
Tamponade
Poor Left Ventricular Output ("Ineffective" Circulating Volume)
Aortic valve stenosis or insufficiency
Myocardial infarction
Mitral valve stenosis or insufficiency
Ischemic cardiomyopathy

Cl-Resistant Group

- Replace K (and ? Mg) deficits
- Block aldosterone action
- Identify basis

Therapy is generally more difficult in patients with ECF volume expansion. If they have renal failure and are very alkalemic ($[H^+]$ <20 nM, pH >7.70) they should receive some H^+ in the form of HCl or NH_4Cl. If these patients are dialyzed, the bath should have the $[HCO_3^-]$ or [acetate] reduced so as to ameliorate rather than contribute to the alkalemia.

If the patient has hyperadrenalism, agents blocking Na reabsorption in the collecting duct (amiloride) or mineralocorticoid antagonists have an important role in the therapy. One must use aggressive K replacement in conjunction with amiloride or spironolactone *with caution*, so as to avoid serious hyperkalemia. In the acute setting, it is safest to order only 1 day of K therapy at a time, seeing the serum $[K^+]$ each day prior to ordering that day's therapy.

Patients with Mg deficiency should have their deficiency corrected, but it is important to follow up on the effectiveness of the therapy, since patients with hyperaldosteronism also have Mg deficiency.

At times, one would like to alleviate the alkalemia quickly (e.g., weaning from a ventilator), and acetazolamide is sometimes used in such a setting, but because a large load of $NaHCO_3$-rich fluid is delivered to the collecting duct, this agent promotes a substantial additional K loss. For this reason, we prefer an intravenous HCl preparation in such patients. With IV HCl, one also has better control over the degree of fall in $[HCO_3^-]$ than with acetazolamide.

Patients with Bartter's syndrome are more difficult to treat, and multiple therapeutic agents are usually required, including nonsteroidal antiinflammatory agents and agents that block Na reabsorption in the collecting duct (amiloride, spironolactone). Finally, additional KCl, NaCl, and $MgSO_4$ supplements are also necessary to achieve a $[K^+]$ and $[HCO_3^-]$ close to the normal range.

Suggested Readings: Metabolic Alkalosis

Berger BE, Cogan MG, and Sebastian A: Reduced glomerular filtration and enhanced bicarbonate reabsorption maintain metabolic alkalosis in humans. Kidney Int 26:205–208, 1984.

Cogan MG, Liu F-Y, Berger B, et al: Metabolic alkalosis. Symposium on acid-base disorders. Med Clin North Am 67:903–914, 1983.

Harrington JT: Metabolic alkalosis. Kidney Int 26:88–97, 1984.

Kassirer JP and Schwartz WB: The response of normal man to selective depletion of hydrochloric acid. Am J Med 40:10–17, 1966.

Kassirer JP, Berkman P, Lawrenz DR, et al: The critical role of chloride in the correction of hypokalemic alkalosis in man. Am J Med 38:172–189, 1965.

Luke RG and Galla JH: Chloride-depletion alkalosis with a normal extracellular fluid volume. Am J Physiol F 244:419–424, 1983.

5

RESPIRATORY ACID-BASE DISTURBANCES

THE BIG PICTURE

Respiratory acid-base disorders are caused by changes in the P_{CO_2}: a rise producing respiratory acidosis and a fall producing respiratory alkalosis.

The chronic respiratory acid-base disorders have larger changes in plasma [HCO_3^-] than the acute ones, owing to changes in renal H^+ secretion.

In respiratory acidosis, the P_{CO_2} and the [H^+] rise. *Acute*: For every mm Hg increase in P_{CO_2} from 40, expect 0.77 nM increase in [H^+] from 40 nM. *Chronic*: For every mm Hg increase in P_{CO_2} from 40, expect 0.32 nM increase in [H^+] from 40 nM.

In respiratory alkalosis, the P_{CO_2} and the [H^+] fall. *Acute*: For every mm Hg fall in P_{CO_2} from 40, expect 0.74 nM fall in [H^+] from 40 nM. *Chronic*: For every mm Hg fall in P_{CO_2} from 40, expect 0.17 nM fall in [H^+] from 40 nM.

In practice, for each acute doubling of the P_{CO_2}, expect the plasma [HCO_3] to rise 2.5 mM. The converse is also true.

From a pulmonary viewpoint, patients with respiratory acidosis are best classified as those who *cannot* breathe and those who *will not* breathe.

There are 3 major reasons why patients hyperventilate and have respiratory alkalosis: (1) central stimulation, (2) peripheral chemoreceptors (hypoxia), and (3) pulmonary reflexes.

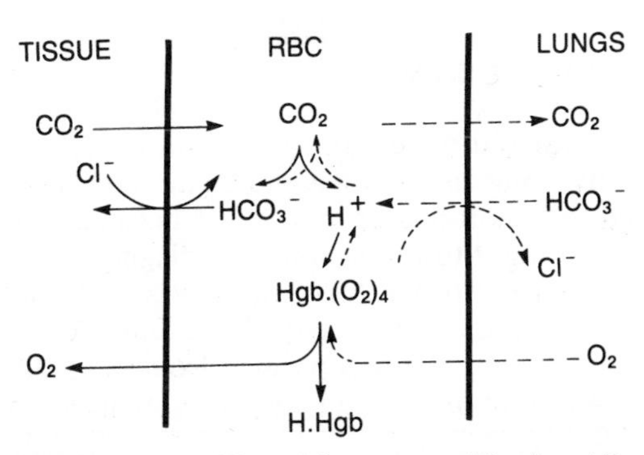

Figure 5–1. Interplay of CO_2 and O_2 transport. The dotted lines indicate the reactions at the level of the lungs. The solid lines indicate the reactions at the level of the tissues (see text for details).

THE PHYSIOLOGY OF O_2 AND CO_2 TRANSPORT

Oxygen Transport. O_2 is transported bound to hemoglobin; 4 O_2 per hemoglobin, and normally there are 2 mmol (140 g) of hemoglobin/L of blood. The affinity of hemoglobin for O_2 is strong but is reduced by H^+, CO_2, and 2,3-diphosphoglycerate.

Effect of H^+ (Bohr Effect). An increase in the [H^+] of plasma enhances the discharge of O_2 from hemoglobin because H^+ binding to hemoglobin lessens hemoglobin's affinity for O_2. This is important during exercise (lactic acid is produced, raising the [H^+], which facilitates O_2 release [Fig. 5–1]).

Effect of CO_2. The increase in P_{CO_2} in tissue facilitates O_2 release from hemoglobin by 2 mechanisms: a rise in [H^+] and the formation of carbamates.

Effect of 2,3-DPG. The negative charges on 2,3-DPG bind to hemoglobin, reducing affinity for O_2. Increases in 2,3-DPG in RBC occur when there is a need for more O_2 delivery. The signal is the alkalemia that results from hyperventilation (e.g., high altitude, exercise, anemia). These effects occur after several hours.

Carbon Dioxide Transport. 10 mmol of CO_2 per minute diffuse into RBC containing carbonic anhydrase, which converts the CO_2 to H^+ and HCO_3^- (Fig. 5–1). The lower P_{CO_2} in the RBC aids further CO_2 diffusion. The HCO_3^- formed (8.5 mmol/min) in the RBC is transported into the plasma in exchange for Cl^- ("chloride shift"), and the H^+ bind to hemoglobin.

Once at the lung, the process is reversed. The lower P_{CO_2} of alveolar air aids in CO_2 diffusion from blood to the alveolus. The high P_{O_2} of alveolar air promotes O_2 binding to hemoglobin, which leads to the dissociation of bound H^+. These H^+ combine with HCO_3^- that were in plasma (reversal of the "chloride shift"). The net result is oxygenation in concert with CO_2 unloading at the lung.

ILLUSTRATIVE CASE

A 64 year old male with a long history of chronic obstructive lung disease on home O_2 therapy developed a cough and shortness of breath. He became confused and was brought to the hospital. A diagnosis of pneumonia and acute respiratory failure was made, and he was intubated and ventilated. The results 2 days later are shown below. He was weaned from the ventilator on day 4 of therapy, remaining on O_2 by mask.

	Previous Results	Admission Values	Treatment	
			Day 2	*Day 4*
$[H^+]$ (nM)	46	63	40	64
pH	7.34	7.20	7.40	7.19
P_{CO_2} (mm Hg)	60	80	40	70
P_{O_2} (mm Hg)	60	35	180	55

Questions (answers on page 123)

1. How would you analyze the change from the previous clinical values to the admission values?
2. What is your evaluation of the blood gases while the patient was on the ventilator on day 2 of therapy?
3. Why are his postextubation values on day 4 of therapy so different from his original values?

DEFINITIONS

RESPIRATORY ACIDOSIS:
INCREASED P_{CO_2} AND PLASMA $[H^+]$
RESPIRATORY ALKALOSIS:
DECREASED P_{CO_2} AND PLASMA $[H^+]$

Respiratory acid-base disorders arise from primary changes in P_{CO_2}. There is a very large flux of CO_2 relative to the concentration of carbon dioxide in the plasma, i.e., 10 mmol of CO_2 are produced per minute, and the P_{CO_2} + H_2CO_3 are only 1.2 mM. Should a discrepancy develop transiently between CO_2 production and its removal by the lungs, the resultant change in arterial P_{CO_2} then displaces

Answers

1. His plasma $[HCO_3^-]$ at the previous clinic visit can be calculated as follows:

$$[HCO_3^-] = 23.9 \times 60 \text{ mm Hg}/46 \text{ nM} = 31 \text{ mM}$$

The $[H^+]$ is that which is expected in simple CHRONIC respiratory acidosis with a P_{CO_2} of 60 mm Hg, i.e., the rise in $[H^+] = 0.32 \times 20$ nM or 6.4 nM (46 nM − 40 nM = 6 nM). When the patient was admitted, he had an acute increase in P_{CO_2} to 80 mm Hg with profound hypoxia. Therefore, the severe acidemia was largely due to the presence of acute respiratory acidosis superimposed on chronic respiratory acidosis. The acute respiratory acidosis was due to pneumonia.

2. While he is being ventilated on day 2 of therapy, his acid-base state has returned to normal: $[H^+] = 40$ nM, $[HCO_3^-] = 24$ mM, $P_{CO_2} = 40$ mm Hg. Although one can achieve a normal acid-base status with artificial ventilation, we know that this man has severe lung disease, and he will not be able to maintain a P_{CO_2} of 40 mm Hg once he is extubated. Therefore, it is inappropriate to ventilate him to this level; he should be maintained with a P_{CO_2} of 60 mm Hg (his chronic steady-state value) to preserve his renal adaptation to chronic respiratory acidosis. In this case, his kidneys have been "fooled" into thinking his lungs are normal, and therefore he excreted the extra HCO_3^-, and his plasma $[HCO_3^-]$ has fallen to 24 mM instead of the 31 mM that was present prior to the acute illness.

3. With extubation on day 4 of therapy, he has become very acidemic as he has gone from a normal acid-base state to acute respiratory acidosis with a P_{CO_2} of 70 mm Hg. The $[H^+]$ is appropriate for acute respiratory acidosis: the rise in $[H^+] = 30$ mm Hg $\times$ 0.77 = 23 nM or a $[H^+]$ of 63 nM. His plasma $[HCO_3^-]$ is $23.9 \times 70/63 = 26$ mM, which has risen appropriately for acute respiratory acidosis.

This severe acidemia could have been avoided by maintaining the P_{CO_2} at 60 mm Hg during mechanical ventilation, which would have maintained the plasma $[HCO_3^-]$ at 31 mM, his chronic level. In that case, when he was extubated, his $[H^+]$ would at least have been $23.9 \times 70/31 = 54$ nM (pH 7.27).

His P_{CO_2} is now slightly higher than his chronic steady state (70 vs 60 mm Hg). This may represent as yet incomplete resolution of the pneumonia or may indicate permanent destruction of lung tissue—time will tell.

the bicarbonate buffer system equation to the RIGHT.

$$CO_2 + H_2O \rightleftharpoons H_2CO_3 \rightleftharpoons H^+ + HCO_3^-$$

CO_2 accumulation results in an increased ECF $[H^+]$ (RESPIRATORY ACIDOSIS), and a fall in P_{CO_2} displaces the equilibrium to the LEFT, resulting in a fall in ECF $[H^+]$—RESPIRATORY ALKALOSIS.

A critical point to grasp is the relative $[HCO_3^-]$ and $[H^+]$ in plasma. The $[HCO_3^-]$ is 10^6-fold > $[H^+]$, but they are formed in equal amounts. Hence, the rise in $[H^+]$ with CO_2 accumulation is relatively large, and acidosis ensues.

> CO_2 EXCRETION = ALVEOLAR VENTILATION × ALV $[CO_2]$
> *Normal:* 10 mmol/min = 6 L/min × 1.7 mmol/L
> *Chronic respiratory acidosis:* 10 mmol/min = 3 L/min × 3.3 mmol/L

It is useful to think of the CO_2 production and removal system as in a way analogous to creatinine production and creatinine clearance. In the steady state, the CO_2 production rate is constant, and the CO_2 removal rate is a function of the alveolar ventilation and the alveolar concentration of CO_2. As the alveolar ventilation falls (cf. GFR), the CO_2 concentration in the alveolar air must rise (cf. serum creatinine) to remove the CO_2 produced daily.

> Chronic respiratory disorders have a renal response causing:
> In acidosis: Increased plasma $[HCO_3]$
> In alkalosis: Decreased plasma $[HCO_3]$

Chronic respiratory acid-base disturbances influence renal H^+ secretion such that a rise in plasma $[HCO_3^-]$ occurs during chronic respiratory acidosis and a fall occurs during chronic respiratory alkalosis. Thus, chronic respiratory acid-base disturbances have a different steady-state plasma $[HCO_3^-]$, and hence $[H^+]$, than do the acute respiratory acid-base disorders (page 128). Therefore, the clinician must clarify on *clinical grounds* whether the acid-base disturbance is acute or chronic in origin.

BUFFERING OF H^+ IN RESPIRATORY ACIDOSIS

1. Buffering occurs almost exclusively in the ICF because:

a. The major ECF buffer is the bicarbonate system, and its effectiveness is compromised with a defect in CO_2 removal.

b. The ICF buffers are, in large part, nonbicarbonate buffers, and therefore are effective buffers of H_2CO_3. Furthermore, the buffer capacity of the ICF is huge relative to the H^+ load. Therefore, a rise in P_{CO_2} leads to H^+ binding in the ICF and to HCO_3 formation (Equations 1 and 2).

$$CO_2 + H_2O \rightleftharpoons H_2CO_3 \rightleftharpoons H^+ + HCO_3^- \quad (1)$$
$$H^+ + B^0 \rightleftharpoons HB^+ \quad (2)$$

$$\text{Sum (Equations 1 + 2): } CO_2 + H_2O + B^0 \rightleftharpoons HB^+ + HCO_2^-$$

2. Surplus ICF HCO_3^- is formed when H_2CO_3 is buffered in the ICF, and some is exported into the ECF, leading to the rise in the ECF $[HCO_3^-]$.

3. *Quantitative analysis:* Given that the ICF buffer capacity is enormous, the quantity of H^+ buffered in the ICF depends on the rise in ICF $[H^+]$. The ICF $[H^+]$ rise is proportional to the rise in P_{CO_2}. Therefore, the rise in $[HCO_3]$ in the ICF, and hence in the ECF, with a P_{CO_2} rise is proportional to the relative rather than the absolute increment in P_{CO_2}. For example, consider a 20 mm Hg rise in P_{CO_2} under 2 circumstances:

A P_{CO_2} rise from 20 to 40 mm Hg = 100% change
A P_{CO_2} rise from 60 to 80 mm Hg = 33% change

In these circumstances a 20 mm Hg increase in P_{CO_2} has a very different impact on ICF $[H^+]$, depending on the original level of the P_{CO_2}; one would anticipate that the larger the percent change in P_{CO_2}, the larger the generation of HCO_3^- from ICF buffers.

4. *Empiric observations:* A P_{CO_2} rise of 2-fold causes a 2.5 mM rise in plasma $[HCO_3]$, and similarly, a halving of the P_{CO_2} causes a 2.5 mM fall in plasma $[HCO_3]$.

THE ALVEOLAR-ARTERIAL O_2 GRADIENT

Alveolar-Arterial P_{O_2} Gradient
- P_{O_2} alveolar − P_{O_2} arterial
- Alveolar P_{O_2} is calculated as: Inspired P_{O_2} − 1.25 × arterial P_{CO_2}

Although respiratory acid-base disorders are defined by changes in the blood P_{CO_2}, important information can also be derived by interpreting the blood P_{O_2}. Specifically, a determination of the O_2 gradient between alveolar air and arterial blood (the A-a gradient) provides useful information. The ALVEOLAR P_{O_2} can be estimated from the abbreviated alveolar gas equation:

$$P_{A_{O2}} = P_{I_{O2}} - P_{CO_2}/RQ = P_{I_{O2}} - 1.25\,(P_{CO_2})$$

($P_{A_{O2}}$ and $P_{I_{O2}}$ are the P_{O_2} of alveolar and inspired air respectively; RQ is the respiratory quotient. See page 127 for more detail.)

The gradient between the alveolar and the arterial P_{O_2} is normally <15 mm Hg. The following questions illustrate the value of the A-a gradient.

Questions (Answers on page 127) (Hint: Look at P_{CO_2} vs $[H^+]$ and the A-a gradient.)

A patient presents in coma following a drug overdose with the following blood gases:

pH	7.24	P_{CO_2}	64 mm Hg
$[H^+]$	58 nM	P_{O_2}	60 mm Hg

4a. The basis of the hypoxia is thought to be aspiration. Is that correct?

Two days later he is awake but coughing. The following blood gases are obtained:

pH	7.60	P_{CO_2}	24 mm Hg
$[H^+]$	25 nM	P_{O_2}	70 mm Hg

4b. As his P_{O_2} has improved, he is declared ready for discharge. Is that an appropriate decision?

THE ALVEOLAR-ARTERIAL (A-a) GRADIENT

The arterial Po_2 is a function of both the Po_2 of alveolar air and the diffusibility of O_2 across the alveolar capillary membrane. Air is approximately 79% nitrogen and 21% O_2. In the lungs, CO_2 and water vapor are part of the nonnitrogen gases. Therefore, if one is breathing room air, as the Pco_2 of alveolar air rises, the Po_2 must fall. Similarly, for any given inspired O_2 tension, a reduction in alveolar Pco_2 (hyperventilation) leads to an increase in alveolar and, hence, arterial Po_2.

The calculation of the A-a gradient allows one to see how much of the derangement in arterial Po_2 is due to a change in alveolar Pco_2 and how much is due to intrinsic lung disease (reduced transfer of O_2 from alveolus to blood). One must have an accurate analysis of the Po_2 of the inspired air to calculate the A-a gradient.

If air is 21% O_2, barometric pressure is 760 mm Hg, and water vapor pressure is 47 mm Hg, the Po_2 of inspired air is 0.21 (760 − 47) = 150 mm Hg.

Answers

4a. Using the alveolar gas equation (as above), begin by calculating the ALVEOLAR Po_2:

The P_IO_2 = 0.21 (760 − 47) = 150 mm Hg, when 47 mm Hg is the water vapor pressure.

The ALVEOLAR Po_2 = (150 − [1.25 × 64]) = 70 mm Hg.

The ARTERIAL Po_2 is 60 mm Hg; therefore, the A-a gradient is 10 mm Hg (70 − 60)—a normal value.

With respect to the acid-base disturbance, the plasma [HCO_3^-] = 23.9 × 64/58 = 26 mM, and the [H^+] is 58 nM; both are expected values for acute respiratory acidosis (see page 128). Together, acute CO_2 retention and hypoxia with a normal A-a gradient suggest that the patient has decreased ventilation, owing to central respiratory suppression from drug overdose rather than intrinsic lung disease. It is *incorrect* to postulate aspiration based on these values.

4b. The alveolar Po_2 in the second set of blood gases is

$$150 - 30 = 120 \text{ mm Hg}$$

Therefore the A-a gradient is 120 − 70 = 50 mm Hg. On this occasion, the patient has a frankly increased A-a gradient, indicating the presence of a disease impairing O_2 exchange between alveoli and blood. The hypoxia is less evident, owing to hyperventilation (if the patient had had a normal Pco_2 and an A-a gradient of 50 mm Hg, the Po_2 would have been 50 mm Hg).

Therefore, there is now significant lung disease present and you must establish its basis.

Respiratory Acidosis

Respiratory acidosis is a result of alveolar hypoventilation. The clinician's problem is to establish the basis of the hypoventilation, which can be divided into 2 groups:

1. Those patients who *will not* breathe (i.e., lack of stimulus).
2. Those patients who *cannot* breathe (i.e., equipment defective).

In addition, patients with a fixed alveolar ventilation (i.e., those on ventilators) develop increased P_{CO_2} if they have increased CO_2 production.

Patients Who Will Not Breathe. These patients lack the normal respiratory neurologic input at 1 of 3 levels:

1. Cerebral nonrhythmic neurones.
2. Brain stem rhythmic function.
3. Upper airway reflexes.

See Table 5–1 for a summary of these disorders and the means for their detection.

Patients Who Cannot Breathe. The causes of inability to breathe fall into 3 categories:

1. Primary muscle disorders.
2. Increased elastic work (restrictive disease).
3. Increased resistance to flow (obstructive disease).

These disorders are described in Table 5–2, page 138.

ACID-BASE ASPECTS OF RESPIRATORY ACIDOSIS (Fig. 5–2, page 138)

ACUTE RESPIRATORY ACIDOSIS:

- FOR EVERY mm Hg INCREASE IN P_{CO_2} FROM 40 mm Hg, EXPECT A 0.77 nM INCREASE IN $[H^+]$ FROM 40 nM
- PLASMA $[HCO_3]$ RISES 2.5 mM WITH A P_{CO_2} DOUBLING

CHRONIC RESPIRATORY ACIDOSIS:

FOR EVERY mm Hg INCREASE IN P_{CO_2} FROM 40 mm Hg, THE $[H^+]$ INCREASES 0.3 nM, AND THE $[HCO_3]$ INCREASES 0.3 mM.

Respiratory acidosis occurs when ventilation transiently

TABLE 5–1. Patients Who Will Not Breathe

Category	Cause	Diagnostic Tests
Cerebral Nonrhythmic Neurones	• POSTHYPOXIC BRAIN DAMAGE • CEREBRAL TRAUMA • INTRACRANIAL DISEASE • PSYCHOTROPIC DRUGS	• CANNOT: COUGH TALK HOLD BREATH
Brain Stem Rhythmic Function	• BRAIN STEM HERNIATION • ENCEPHALITIS • CENTRAL SLEEP APNEA • METABOLIC ALKALOSIS (SEVERE) • DRUGS (SEDATIVES, NARCOTICS)	• ABNORMAL RESPONSE TO CO_2 AND O_2
Upper Airway Reflexes	• BULBAR PALSY • ANTERIOR HORN CELL LESIONS • DISRUPTION OF AIRWAY	• CANNOT SWALLOW • ABSENT NASAL, TRACHEAL, AND PHARYNGEAL REFLEXES

TABLE 5–2. Patients Who Cannot Breathe

Category	Cause	Diagnostic Tests
Respiratory Muscle Disease	• MYASTHENIA • RELAXANT DRUGS • MUSCULAR DYSTROPHY • MUSCLE FATIGUE	• REDUCED MAXIMUM INSPIRATORY AND EXPIRATORY PRESSURES
Increased Elastic Work	• INTERSTITIAL LUNG DISEASE • PULMONARY FIBROSIS	• REDUCED LUNG VOLUMES • REDUCED LUNG COMPLIANCE • NORMAL FLOW RATES
Increased Resistance to Flow	• ASTHMA • BRONCHITIS • EMPHYSEMA • UPPER AIRWAY OBSTRUCTION	• REDUCED FLOW RATES • INCREASED AIRWAY RESISTANCE

fails to remove the CO_2 produced by normal metabolism. As a result, the alveolar P_{CO_2} rises, increasing the arterial P_{CO_2}. At this new arterial and alveolar P_{CO_2}, the CO_2 produced can now be removed despite the reduced ventilation (see page 124).

Acute Respiratory Acidosis. With acute CO_2 retention, there is an increase in $[HCO_3^-]$ of 2.5 mM for each 2-fold rise in P_{CO_2}. The increase in plasma $[H^+]$ is minimized by the intracellular buffering of most of the H^+ load (page 125). Thus, the patient with acute CO_2 retention has ACIDEMIA, an elevated P_{CO_2}, and a slight rise in the plasma $[HCO_3^-]$ (the degree of elevation in $[H^+]$ is relatively much greater than the elevation in $[HCO_3^-]$, owing to the *very* low normal plasma $[H^+]$).

In acute respiratory acidosis in previously normal patients, an empirical linear relationship has been found between the $[H^+]$ and the P_{CO_2} (see page 138), so that FOR EVERY mm Hg increase in P_{CO_2}, one EXPECTS a 0.77 nM INCREASE in $[H^+]$. In patients who do not begin with a normal acid-base state (e.g., coexisting metabolic acidosis or alkalosis), the plasma $[HCO_3]$ rises 2.5 mM for a 2-fold change in P_{CO_2} (see page 125 for more detail).

It is important to relate the patient's P_{CO_2} to the acid-base state for another reason: whereas the normal P_{CO_2} at a $[HCO_3^-]$ of 25 mM is 40 mm Hg, the expected P_{CO_2} in metabolic acidosis at a plasma $[HCO_3]$ of 10 mM is 25 rather than 40 mm Hg (see page 44). Therefore, at a $[HCO_3]$ of 10 mM, a patient with a P_{CO_2} of 40 mm Hg has respiratory acidosis. This P_{CO_2} difference is very important to the plasma $[H^+]$; in the above case, the [H] is 60 nM with a P_{CO_2} of 25, and 96 nM at a P_{CO_2} of 40 mm Hg.

Chronic Respiratory Acidosis. In chronic respiratory acidosis, renal HCO_3 handling is altered, and the ECF $[HCO_3^-]$ rises. Empirically, the near linear relationship between the plasma $[H^+]$ and P_{CO_2} has a slope of 0.32 (for every mm Hg increase in P_{CO_2}, there is a 0.32 nM increase in $[H^+]$; for every 10 mm Hg increase in P_{CO_2}, there is a 3 mM increase in the plasma $[HCO_3]$.

Diagnostic Approach to Respiratory Acidosis. The diagnostic approach to respiratory acidosis is outlined in Flow chart 5–A, page 131. First, decide whether the patient has chronic lung disease by the history, physical exam, and records. Then compare the acid-base status with that expected for that acid-base disorder. If a discrepancy exists, a mixed disorder is present.

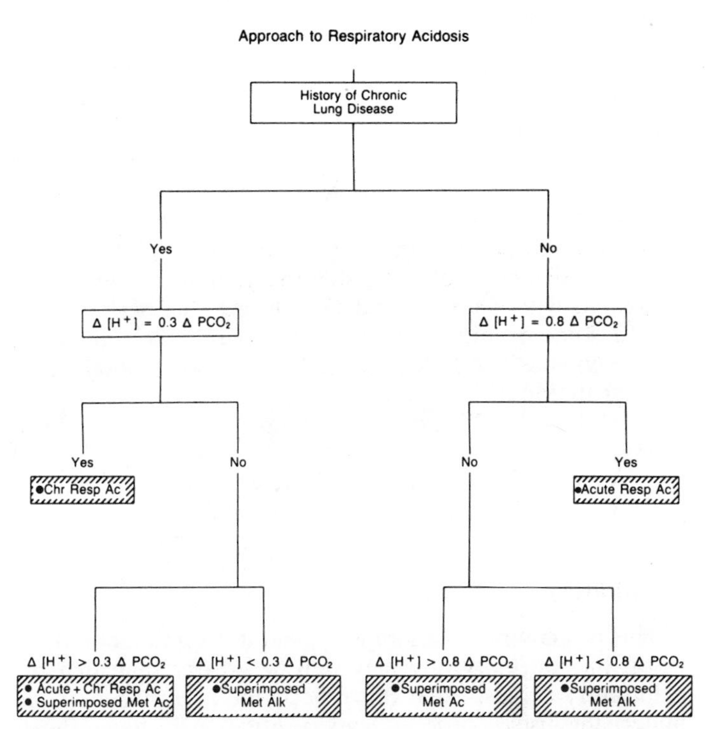

Flow Chart 5–A. Diagnostic approach to respiratory acidosis. In a patient with an elevated P_{CO_2}, begin with the *clinical* examination for the presence of chronic lung disease. If there is no evidence of chronic lung disease, the patient is assumed to have acute respiratory acidosis. If the patient is more acidemic than expected, there is a superimposed metabolic acidosis; if more alkalemic, a superimposed metabolic alkalosis. On the other hand, if there is a history of chronic lung disease and the patient is more acidemic than predicted, he or she may have a component of either acute respiratory acidosis or metabolic acidosis. If the patient is more alkalemic, there is a superimposed metabolic alkalosis.

Questions (Answers on page 133)

5. Three patients all experience an acute increase in P_{CO_2} to 80 mm Hg. Each has a different $[H^+]$ (the pH is shown in parentheses). What is the acid-base status in each case?

$[H^+]$:

A. 96 nM (7.02)
B. 70 nM (7.15)
C. 50 nM (7.30)

6. A patient with chronic stable obstructive lung disease (P_{CO_2} = 60 mm Hg, plasma $[HCO_3^-]$ = 31 mM) developed shortness of breath and was admitted with a diagnosis of congestive heart failure. He was treated with diuretics and O_2. During this period he had some vomiting, but his chest x-ray showed improvement. The following blood gases were obtained after therapy:

$[H^+]$ = 38 nM, P_{CO_2} = 80 mm Hg, P_{O_2} = 70 mm Hg

A. What is the basis of the increased CO_2 retention?
B. What therapy would you institute?

Respiratory Alkalosis

This is a common abnormality, and it is often ignored. In truth, the mortality rate may well be greater than that for respiratory acidosis, and this reflects the importance of the underlying disease process. Hyperventilation is often difficult to recognize clinically, and the diagnosis is often made only by blood gas determination.

Respiratory alkalosis occurs when the ventilatory removal of CO_2 transiently exceeds its rate of production: thus, the alveolar P_{CO_2} and the arterial P_{CO_2} fall. At this lower level of arterial P_{CO_2}, the daily CO_2 production is then removed by increased ventilation, leading to a new steady state (see section on CO_2 excretion, page 124).

Respiratory alkalosis may result from stimulation of the peripheral chemoreceptors (hypoxia or hypotension), the afferent pulmonary reflexes (intrinsic pulmonary disease), or central stimulation by a host of stimuli (Table 5–3, page 137).

Answers

5. The $[HCO_3^-]$ can be calculated in each case from the Henderson equation (page 4).

Patient	A	B	C
$[H^+]$ (nM)	96	70	50
$[HCO_3^-]$ (mM)	20	27	38

Since each patient has acute respiratory acidosis, we would expect the plasma $[HCO_3^-]$ to increase 2.5 mM in association with the acute doubling of the P_{CO_2} (page 128). In fact, Patient A has a $[HCO_3^-]$ that is BELOW normal rather than increased; therefore, Patient A has METABOLIC ACIDOSIS in addition to the acute respiratory acidosis.

In Patient B, the rise in plasma $[HCO_3^-]$ is close to 2.5 mM, so this could be simple acute respiratory acidosis.

In Patient C, the plasma $[HCO_3^-]$ is 38 mM and therefore was 35.5 prior to the acute increase in P_{CO_2} from 40 to 80 mm Hg (an increase of 2.5 mM for each doubling of the P_{CO_2}). Thus, Patient C has metabolic alkalosis in addition to the acute respiratory acidosis (see Chapter 4).

Alternatively, we can calculate the expected $[H^+]$ for a patient with an acute increase in blood P_{CO_2} from 40 to 80 mm Hg on the basis of $[H^+] = 0.77 \times P_{CO_2}$; the $[H^+]$ is $40 + 31 = 71$, with a $[HCO_3^-] = 23.9 \times 80/71 = 27$ mM—virtually identical to Patient B. Thus, Patient B indeed has blood gas values compatible with acute respiratory acidosis, and both approaches are consistent.

6a. The patient's chronic steady-state $[H^+] = (23.9 \times 60/30 =$ 46 nM, which is appropriate for chronic respiratory acidosis (rise in $[H^+] = 6$ nM $= 0.32 \times 20$). Following therapy, his P_{CO_2} has risen, but his $[H^+]$ has fallen to 38 nM and the $[HCO_3]$ is thus $23.9 \times 80/38 = 50$ mM. This $[HCO_3^-]$ is much higher than it should be for chronic respiratory acidosis. Thus, the patient has a mixed disorder: chronic respiratory acidosis and metabolic alkalosis. The metabolic alkalosis is due to both the vomiting and the diuretic therapy (see Chapter 4). Metabolic alkalosis does remove the acidemic stimulus to ventilation and is associated with worsening of CO_2 retention in patients with chronic respiratory acidosis. Usually the degree of hypoventilation induced by metabolic alkalosis is limited by the development of hypoxia; however, in this case, because the patient was receiving O_2, a very significant degree of hypoventilation ensued.

6b. The appropriate therapy is to correct the metabolic alkalosis, which is of the saline-responsive variety. The patient will most likely be K-depleted as well (ask for additional lab data). Presumably the patient is now free of congestive heart failure and can be given KCl and some NaCl with caution.

THE ACID-BASE ASPECTS OF RESPIRATORY ALKALOSIS (Fig. 5–2, page 138)

- IN ACUTE RESPIRATORY ALKALOSIS, FOR EVERY mm Hg DECREASE IN P_{CO_2} FROM 40 mm Hg, EXPECT A 0.74 nM DECREASE IN [H^+] FROM 40 nM
- IN CHRONIC RESPIRATORY ALKALOSIS, FOR EVERY mm Hg DECREASE IN P_{CO_2} FROM 40 mm Hg, EXPECT A 0.17 nM DECREASE IN [H^+] FROM 40 nM

The acid-base impact of respiratory alkalosis is analogous to respiratory acidosis in that the acute and chronic states differ, owing to the role played by the kidney.

Acute Respiratory Alkalosis. In acute respiratory alkalosis, there is displacement of the HCO_3^- buffer system to the left, leading to a reduction in both the [H^+] and the [HCO_3^-].

$$CO_2 + H_2O \rightleftharpoons H_2CO_3 \rightleftharpoons H^+ + HCO_3$$

The impact on the ECF [H^+] in acute respiratory alkalosis is virtually identical to that in acute respiratory acidosis (although opposite in direction). For every mm Hg reduction in the P_{CO_2}, you should see a 0.74 nM reduction in [H^+].

Chronic Respiratory Alkalosis. In chronic respiratory alkalosis, there is a temporary small suppression of renal NH_4^+ production and excretion, and the ECF [HCO_3^-] falls (continued HCO_3^- consumption due to neutralization of dietary H^+ without equivalent renal new HCO_3^- formation) until the plasma [H^+] approaches normal. This is the only acid-base disorder in which a normal plasma [H^+] is expected. Of course, there can also be more than one disturbance present, i.e., the coexistence of metabolic acidosis and acute respiratory alkalosis, in which the [H^+] may be normal, increased, or decreased, depending on the relative magnitude of the two processes; hence, the differentiation between ACIDEMIA and ACIDOSIS; see Answer 7.

Diagnostic Approach to Respiratory Alkalosis. See Flow Chart 5–B, page 135.

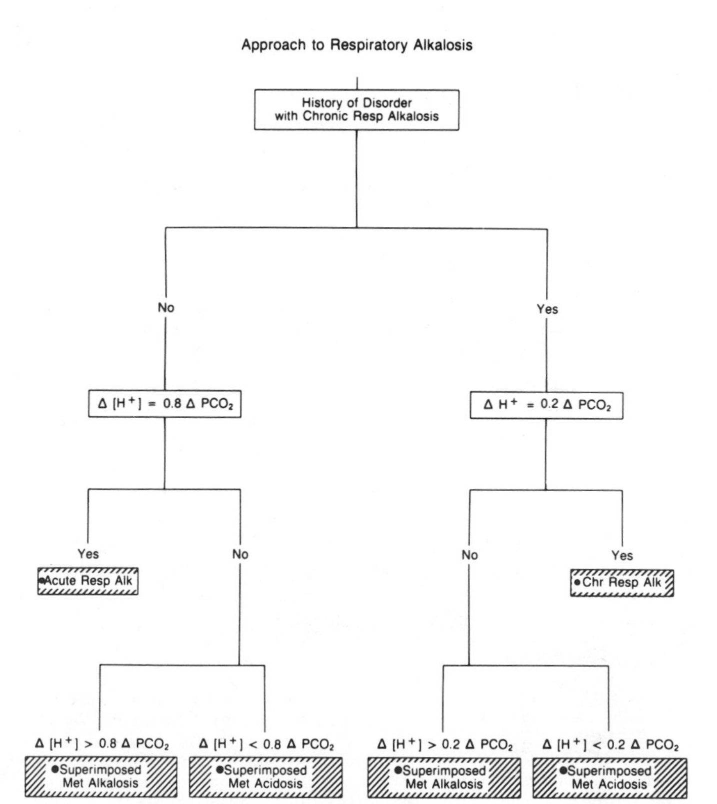

Flow Chart 5–B. Respiratory alkalosis. The parameters to evaluate are shown in the open boxes, and the final diagnoses are shown in the shaded boxes. The Δ sign represents a decrease in H^+ or P_{CO_2}. In a patient with a low P_{CO_2}, begin by looking for a disorder associated with chronic respiratory alkalosis (Table 5–3, page 137). If one is present, the patient is presumed to have chronic respiratory alkalosis and is assessed accordingly. On the other hand, if a chronic disorder is absent, the patient is assumed to have acute respiratory alkalosis (see text for details).

Questions (Answers on page 137)

7. A 30 year old businessman having just returned from Europe suddenly developed a severe left-sided pleuritic chest pain and hemoptysis. He had no history of chest disease and exercised regularly. He was cyanotic, had an elevated jugular venous pressure (8 cm above the sternal angle), and the blood pressure was 80/50 mm Hg. The blood gases were $[H^+]$ = 40 nM (pH 7.40), P_{CO_2} = 25 mm Hg, and P_{O_2} = 50 mm Hg. What is your analysis of the blood gases and your clinical diagnosis?

8. A patient with cirrhosis of the liver was found in a confused state by his landlady. His physical examination was normal except for a low blood pressure and the stigmata of chronic liver disease. His laboratory results were as follows:

Na	133 mM	$[H^+]$	36 nM	(pH 7.44)
K	3.3 mM	pCO_2	20 mm Hg	
Cl	115 mM	$[HCO_3^-]$	13 mM	

His initial diagnosis was lactic acidosis secondary to severe hepatic insufficiency. Do you agree with this diagnosis? If not, why, and what is your diagnosis?

Suggested Reading

Arbus GS, Hebert LA, Levesque PR, et al: Characterization and clinical application of the "significance band" for acute respiratory alkalosis. N Engl J Med 280:117–123, 1969.

Bercovici M, Chen CB, Goldstein MB, et al: Effect of acute changes in the Pa_{CO_2} on acid-base parameters in normal dogs and dogs with metabolic acidosis or alkalosis. Can J Physiol Pharmacol 61:1166–1173, 1983.

Brackett NC, Cohen JJ, and Schwartz WB: Carbon dioxide titration curve of normal man. N Engl J Med 272:6–12, 1965.

Gennari FJ, Goldstein MB, and Schwartz WB: The nature of the renal adaptation to chronic hypocapnia. J Clin Invest, 51:1722–1730, 1972.

Madias NE, Adrogue HJ, Cohen JJ, et al: Effect of natural variations in Pa_{CO_2} on plasma $[HCO_3]$ in dogs: a redefinition of normal. Am J Physiol 235:F30–F35, 1979.

Madias NE, Wolf CJ, and Cohen JJ: Regulation of acid-base equilibrium in chronic hypercapnia. Kidney Int 27:538–543, 1985.

Rebuck AS, Slutsky AS: Control of breathing in diseases of the respiratory tract and lungs. Handbook of Physiological Society, Section 3, Part 2, pp 771–791, 1986.

Schwartz WB, Brackett NC, Cohen JJ: The response of extracellular hydrogen ion concentration to graded degrees of chronic hypercapnia: The physiologic limits of the defense of pH. J Clin Invest 44:291–301, 1965.

TABLE 5–3. Causes of Respiratory Alkalosis

Hypoxia
INTRINSIC PULMONARY DISEASE, HIGH ALTITUDE, CONGESTIVE HEART FAILURE, CONGENITAL HEART DISEASE (CYANOTIC)
Pulmonary Receptor Stimulation
PNEUMONIA, PULMONARY EMBOLISM, ASTHMA, PULMONARY FIBROSIS
Drugs
SALICYLATES ARE THE MOST COMMON. OTHERS INCLUDE NIKETHAMIDE, CATECHOLAMINES,THEOPHYLLINE, PROGESTERONE
CNS Disorders
SUBARACHNOID HEMORRHAGE, CHEYNE-STOKES RESPIRATION, PRIMARY HYPERVENTILATION SYNDROME
Miscellaneous
PSYCHOGENIC HYPERVENTILATION, CIRRHOSIS, FEVER, GRAM-NEGATIVE SEPSIS, FOLLOWING CORRECTION OF METABOLIC ACIDOSIS

Answers

7. From the patient's history, there is no obvious chronic stimulus for hyperventilation; this suggests that respiratory alkalosis might be acute. The plasma $[HCO_3^-]$ is 15 mM ($23.9 \times 25/40$). Therefore, if the patient has acute respiratory alkalosis with a P_{CO_2} of 25 mm Hg, his plasma $[H^+]$ should be 29.5 nM ($40 - [0.7 \times 15]$), and plasma $[HCO_3^-]$ should be 20 mM ($23.9 \times 25/29.5$). Since his plasma $[H^+]$ is higher and his plasma $[HCO_3^-]$ lower, he has metabolic acidosis (lactic acidosis due to low oxygen delivery to tissues). Thus, the combination of respiratory alkalosis and metabolic acidosis resulted in a normal plasma $[H^+]$.

8. No; the diagnosis of simple lactic acidosis is incorrect. Although it is true that the plasma $[HCO_3^-]$ is low, the patient is not acidemic, and the plasma AG is not wide (in fact it is narrow [5 mEq/L], presumably because of hypoalbuminemia due to liver failure). This leaves little room for the 12 mM lactate that would be present if this patient had lactic acidosis.

Starting with his plasma $[H^+]$, it is in the normal range, suggesting that he has a mixed acid-base disturbance (Chapter 2, page 31). However, chronic respiratory alkalosis can return the $[H^+]$ to normal; therefore, this is a consideration, as he is hyperventilating ($P_{CO_2} = 20$ mm Hg). The expected $[H^+]$ for someone with a chronic P_{CO_2} of 20 is $40 - P_{CO_2} \times 0.17 = 40 - 3.4 = 36$ nM—exactly what we saw. Thus, this patient's acid-base status is in keeping with a diagnosis of chronic respiratory alkalosis, despite the fact that the plasma $[HCO_3^-]$ is only 13 mM.

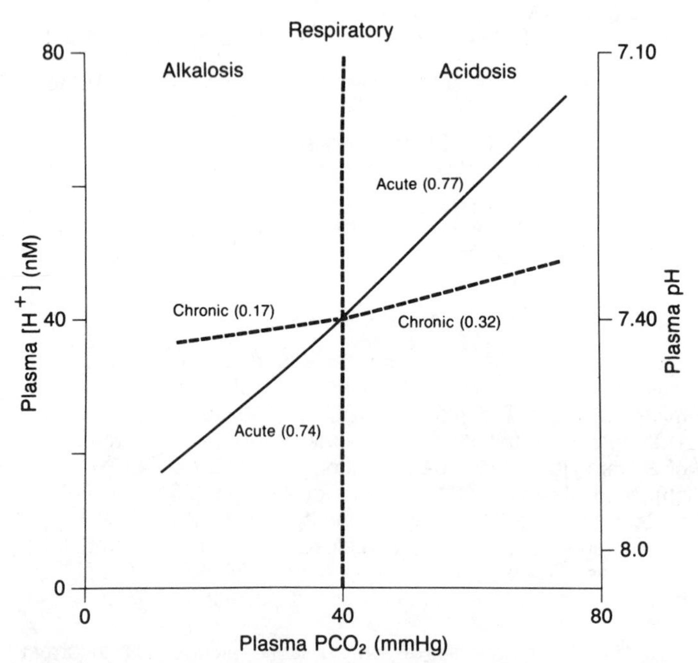

Figure 5–2. The linear relationship between $[H^+]$ and P_{CO_2} in acute and chronic respiratory acid base disturbances. The slopes indicated on the figure enable the physician to predict the $[H^+]$ for any acute or chronic increase or decrease in P_{CO_2}.

6

SODIUM AND WATER PHYSIOLOGY

THE BIG PICTURE: Body Compartments

Approximately 60% of body weight is water; ⅔ of this water is intracellular fluid (ICF) and ⅓ is extracellular fluid (ECF). In a 70 kg person, total body water is 42 L, 28 L ICF and 14 L ECF.

Water crosses cell membranes rapidly; therefore, the osmolality in the ECF and the ICF is equal.

Particles that readily cross the cell membrane and achieve an equal concentration in the ECF and the ICF (e.g., urea and ethanol) can be ignored with respect to their influencing water distribution during steady state.

The major function of Na^+ (and anions Cl^- and HCO_3^-) is to keep water out of cells, thereby maintaining ECF volume (glucose acts like Na for some cells only).

The particles that determine ICF volume are large macromolecular anions and their attendant cations (K^+); these anions rarely leave the ICF.

The ECF *[Na]* (Na/H_2O) reflects the ICF volume of most cells; hypernatremia signals ICF contraction, and

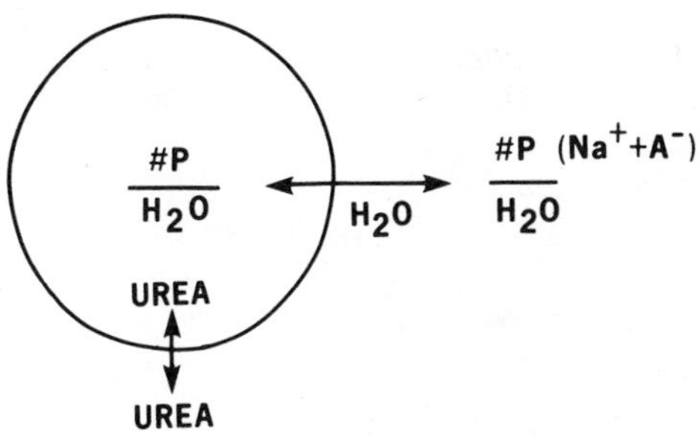

Figure 6–1. Review of factors regulating water distribution across cell membranes. The circle repesents the cell membrane. Water crosses this membrane very rapidly, and osmotic equilibrium is achieved. Particles such as urea (and alcohol) also cross the membrane rapidly, and their concentrations are equal in the ICF and the ECF; hence, they play no role in water distribution. The major particles (P), restricted largely to the ECF, are Na^+ and the anions Cl^- plus HCO_3^-; the particles restricted primarily to the ICF are predominantly organic phosphate anions together with the major ICF cation, K^+ (for details, see text).

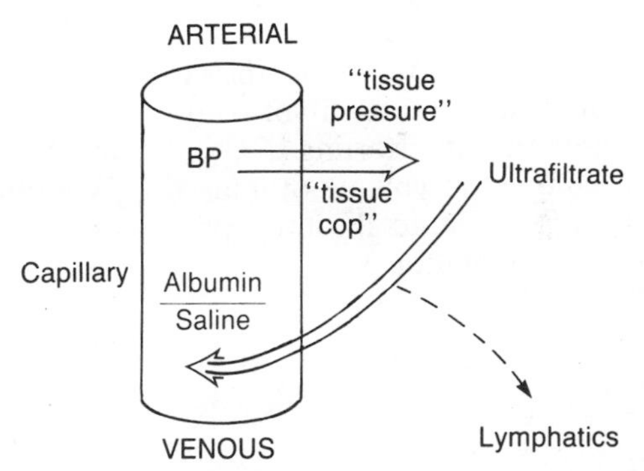

Figure 6–2. Review of factors controlling ECF distribution across the capillary membrane. There are 2 major forces to consider: a higher hydrostatic pressure (BP) causes fluid to leave the vascular space, and the colloid osmotic pressure (COP, albumin/saline) causes fluid to enter the vascular space. Despite the above fluid redistribution, there is *no change* in the ICF volume (the Na/H_2O ratio does not change). Since the interstitial fluid volume is so much larger than the vascular volume, any time you detect expansion (edema, ascites) of the interstitial space, the patient will always have ECF volume expansion, even if the vascular volume is reduced (i.e., *chronic* hypoalbuminemia).

hyponatremia signals ICF expansion, except during hyperglycemia. The total *Na content* (not the Na/H_2O) determines the ECF volume.

An ultrafiltrate (not just water) crosses the capillary wall—a hydrostatic force (blood pressure) is the major outward-driving force, and vascular colloid osmotic pressure (albumin:saline ratio) is the major inward-driving force.

Questions (Answers on page 143)

1. What fluid crosses cell membranes?
2. What fluid traverses the capillary membrane?
3. What particle(s) determine ICF and ECF volumes?
4. What acute ECF and ICF volume changes occur when albumin is abruptly lost?
5. What IV solution would you give if you wanted *all* of it to stay in the ECF? What IV solution would you give to expand the ECF but contract the ICF volume?
6. What proportion of an infusion of dextrose in water ends up in the ICF at steady state?
7. A 70-kg person perspires and loses 4 L of a NaCl solution; the [Na] of this fluid is ½ that in the ECF. Describe in quantitative terms the change in ICF and ECF volumes and the resultant [Na] (for this calculation, assume that the starting [Na] is 150 mM).
8. Can the entire volume of infused D_5W end up in the ICF? (This should illustrate principles governing fluid compartmentation.)

COMPOSITION OF BODY FLUIDS

WATER IS 60% OF MASS, ⅔ ICF AND ⅓ ECF.

The most abundant constituent of the body is water; it accounts for approximately 60% of the body mass (Table 6–1, page 143). This water is divided into two main compartments, ECF water and ICF water. The ECF consists of the plasma water (4% body weight) and interstitial water

TABLE 6–1. Distribution of Body Water

Compartment	% Body Weight	Volume (L)/70 kg
ECF	20*	14
Plasma Water	4	2.8
Interstitial Water	16	11.2
ICF	40	28
Total		
Male (70 kg)	60	42
Female (70 kg)	50	35

*For simplicity, we assume that the ECF volume is 33% of total water. However, the markers used to determine ECF volume are not ideal, and ECF volume could be 27–45% of water, depending on the marker used.

Answers

1. Water (not a salt solution).
2. An ultrafiltrate, e.g., water plus ions and metabolites.
3. *ICF*: K^+ "held in" the ICF by the macromolecular anions. *ECF*: Na^+ (plus Cl^- and HCO_3^-).
4. There is no change in the ECF or ICF volumes. There is a redistribution of the ECF volume: interstitial volume rises and vascular volume decreases.
5. a) Isotonic saline;
   b) hypertonic saline.
6. Two-thirds of the volume, *once the glucose is metabolized*.
7. For simplicity, divide the fluid loss into 2 components: isotonic saline and pure water. Since the [Na] in sweat is ½ of normal in this example, the 4 L could be thought of as 2 L of water and 2 L of isotonic saline. The water loss occurs from the ICF and ECF in direct proportion to their existing volumes; thus ⅔ of the 2 L of water is lost from the ICF (1.33 L) and ⅓ from the ECF (0.67 L). In contrast, the 2 L isotonic saline is lost exclusively from the ECF (no change in the Na/H_2O, the driving force for a water shift). Thus, the ECF is 14 − 0.67 L − 2 L or 11.33 L. Although we ignored the fact that the ECF volume was no longer half that of the ICF volume (a small error), we prefer this form of deductive calculation to the application of a formula. The final [Na] is the Na:water ratio in the ECF. We know the ECF volume (11.33 L from above); the total Na is the original [Na] (150 mmol/L × 14 L) or 2100 mmol; from this, the patient lost 300 mmol of Na; therefore, the total Na content is now 2100 − 300 or 1800 mmol. Therefore, the resulting [Na] is 1800 mmol/11.33 L or almost 159 mM. *Hypernatremia signals a loss of ICF volume in virtually every case*.
8. Normally, ⅔ of a L of D_5W goes into the ICF. However, if something is done in addition, i.e., decreasing the ECF particles (Na,Cl) or increasing the ICF particles, the entire 1 L of infused water will enter the cells.

(16% body weight, i.e., water in tissues between the cells). Water in the abdominal cavity (ascites) or thoracic cavity (pleural effusion) is a component of the interstitial space. In certain disease states, ECF water accumulates to an appreciable degree, out of proportion to the ICF.

To relate total body water to body weight, one presumes that the relative proportion of fat is constant (obviously, this is not true, and a correction must be made for body composition because neutral fat does not dissolve in water). Thus, females tend to have a lower water content per body mass (50% body weight vs 60% for males). Older patients also tend to have a relatively smaller water content due to loss of muscle mass, whereas infants have a higher proportion of water (70%).

Questions (Answers on page 145)

9. Are gastrointestinal secretions ECF or ICF? What is "third-space" fluid?
10. Will exchanging an ICF K^+ for Na^+ have a direct effect on the ICF volume?
11. What is the effect of hyperglycemia on a water shift between the ICF and ECF? (For simplicity, assume that no insulin is present.)

Electrolytes Determining Water Volumes

- Na^+ (and Cl^- plus HCO_3^-) DETERMINE ECF VOLUME
- LARGE MACROMOLECULAR ANIONS PLUS ATTENDANT K^+ DETERMINE ICF VOLUME

Water readily crosses cell membranes (through pores) and rapidly achieves osmotic equilibrium (particle:water ratio). However, not all materials dissolved in water disperse equally in the ICF and ECF (there are differences in permeability, transporters, and active pumps that regulate their distribution).

Distribution of Water Across Cell Membranes (see page 145 for a discussion of osmolality). Water distribution depends on particles restricted to the ICF or to the ECF. These particles account for the "effective osmolality" or tonicity in these compartments.

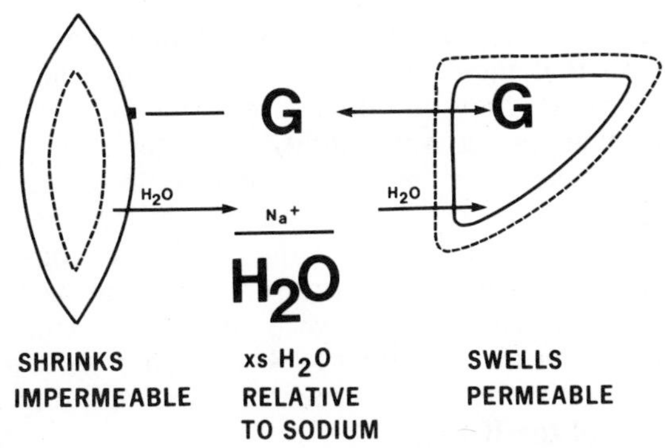

Figure 6–3. Effects of glucose on water shifts. A liver cell is shown on the right-hand side and a muscle cell on the left. The liver always has a [glucose] equal to that in the ECF; hence, glucose is a particle "like urea," with respect to water shifts in this organ. In contrast, the [glucose] in the ECF is always much higher than that in the muscle. Therefore glucose is an "Na-like" particle, with respect to water shifts in this organ. The dotted line reflects the new ICF volumes under the influence of severe hyperglycemia (see also Chapter 12) xs = excess.

Answers

9. The simplest answer is that fluid in the lumen of the GI tract is not, strictly speaking, inside the body. It is an open tube running through the body, and the luminal contents are "outside" body compartments. In contrast to urine and sweat, these GI tract fluids *can be* reabsorbed and, therefore, may only be temporarily outside the ECF and the ICF. Some people may call this a third space for fluid (nonECF, nonICF) (some refer to ascites as third-space fluid, but in truth ascites is indeed part of the ECF.)

10. Since there is no change in the number of particles in the ICF or ECF, there will not be a water shift.

11. Hyperglycemia causes the ECF osmolality to rise. Muscle cells *shrink*; in contrast, no net water movement occurs in liver, owing directly to hyperglycemia. However, the hyponatremia caused by water movement from muscle leads to water entry into the liver cells, and they should *swell* (Fig. 6–3).

FACTS CONCERNING OSMOLALITY AND TONICITY

Osmolality is the force exerted by the number of particles per volume. Particles differ with respect to their distribution in body water. Some, like urea, cross cell membranes and have an equal concentration in all compartments and play no role in the regulation of water distribution. They are "ineffective" osmoles and do not contribute to body *tonicity* (osmoles restricted to one compartment determine compartment volumes).

Particles like urea or alcohol cross cell membranes rapidly enough so that their concentrations are equal in the ICF and ECF; thus, they do not induce water movement. The particles that attract water into cells differ from one cell type to another. The major factor responsible for water movement into the cell is the retention of large macromolecular anions inside the cell. Although the macromolecules do not exert a large osmotic pressure (there are not a large number of particles), they have a large number of cations associated with them (primarily K^+) that not only provide electroneutrality but also account for the majority of the osmolality of the ICF.

- WATER (WITHOUT Na^+) CROSSES THE CELL MEMBRANE UNTIL THE OSMOLALITY (PARTICLE/H_2O RATIO) IS EQUAL ON BOTH SIDES OF THAT MEMBRANE
- TONICITY (EFFECTIVE OSMOLALITY) = TOTAL OSMOLALITY − (mM UREA + mM ALCOHOL)
- ICF PARTICLES RARELY CHANGE
- Na CONTENT DETERMINES THE ECF VOLUME, AND THE ECF [Na] REFLECTS THE ICF VOLUME

Since the ICF macromolecules are largely organic phosphate esters (ATP, creatine phosphate, RNA, DNA, phospholipids, etc.) and are essential for cell function, there are only small net changes in their content. Thus, it follows that the total of ICF particles is relatively "fixed" in number and charge; hence, changes in the particle:water ratio in the ICF usually only come about by a change in the ICF water content (for additional discussion on defense of cell volume, see page 147).

The particles restricted to the ECF (which thereby control the ECF volume) are Na^+ and its attendant anions (Cl^- plus HCO_3^-). It is not that Na^+ cannot cross the cell membrane (although Na^+ permeability is relatively low compared to K^+); rather, any additional Na ions that enter the ICF are actively transported out of the cell by the membrane enzyme, the NaKATPase. Since, under normal conditions, there are roughly twice as many ICF particles as ECF particles, it follows that the ICF volume is 2-fold larger than the ECF volume.

DEFENSE OF CELL VOLUME

The number of particles in the ICF is relatively constant for the majority of cells; however, a minority of cells can specifically defend against a large water shift by varying the number of their ICF particles. This is advantageous for the brain because it is contained in a rigid box (the skull); if brain cells increased in size, they might occupy too large a space in the skull and diminish the cerebral blood supply. In contrast, diminishing the brain ICF volume would shrink the brain, stretching vascular connections to the calvarium, ultimately leading to an intracranial hemorrhage. Recall that the large intracellular macromolecular anions are essential compounds and not expendable—hence, defense of cell volume occurs only if ions or nonelectrolytes are present in abundance and can be translocated across cell membranes.

Volume Regulatory Decrease (controlled return of swollen cells toward their original volume). The mechanism used is primarily the extrusion of electrolytes. Typically, it involves a ouabain-insensitive decrease in ICF K content. Cells vary in their ICF [Cl], the major anion to be lost with K, i.e., the ICF [Cl] is close to 3 mM in muscle and 70 mM in red blood cells. In addition, the time course for this effect varies to a great extent among species and cell types. For example, human B lymphocytes achieve rapid volume regulatory decrease by KCl loss mediated by an increase in cytoplasmic free Ca^{++}. In contrast, T lymphocytes show little change in K permeability in hypotonic media. Amino acids or small peptides, if present in large enough concentrations, may be extruded from cells as part of the volume regulatory decrease (this form of volume regulation is most common in invertebrates). Alternatively, it is possible that an ICF volume decrease could occur without ion extrusion if ions were bound and thus made "osmotically inactive."

Volume Regulatory Increase (controlled return of shrunken cells toward their original volume). The mechanism for a gain in ICF volume during hypernatremia also varies among species and cell types. Typically, it involves a ouabain-insensitive Na influx. In certain cases, the furosemide-sensitive Na, K, 2Cl uniporter is involved, whereas in others, it is the amiloride-sensitive Na^+/H^+ antiporter. It appears that the response of brain cells is slower when hypernatremia rather than hyperglycemia is the cause for hypertonicity.

Factors Controlling the Distribution of Fluid Across the Capillary Membrane (Figure 6–2, page 141):

- MOVEMENT OF AN ULTRAFILTRATE OF PLASMA ACROSS CAPILLARY MEMBRANES DOES NOT CAUSE WATER TO SHIFT BETWEEN ECF AND ICF
- HYDROSTATIC PRESSURE IS THE MAJOR FORCE MOVING FLUID OUT, WHEREAS THE PLASMA [ALBUMIN] IS THE MAJOR FORCE MOVING FLUID INTO THE CAPILLARY LUMEN

Since the capillary wall does not restrict Na movement, an ultrafiltrate of plasma moves across the capillary membrane (contrast this to pure water movement across the cell membrane). Since there is virtually no change in the Na^+:water ratio in the interchange between the vascular and interstitial compartments, these movements do not influence water distribution across the cell membrane.

The factors controlling ultrafiltrate movement across the capillary membrane are shown in Figure 6–2, page 141. The major outward driving force is the hydrostatic pressure gradient. Hence, more ultrafiltrate exists when this hydrostatic pressure gradient increases, i.e., with hypertension and/or a reduction in arteriolar vascular resistance (this resistance varies from moment to moment; hence, prolonged vasoconstriction or vasodilatation can have a major influence on transfer of ultrafiltrate). The hydrostatic pressure at the venous end of the capillary increases with venous hypertension (e.g., with venous obstruction, congestive heart failure).

The major inward (interstitial to vascular) flow of fluid is driven by the colloid osmotic pressure gradient, which is ultimately due to the larger concentration of albumin in the vascular fluid as compared with that in the interstitial fluid (see page 149 for more information). Interstitial fluid is also returned to the venous system via the lymphatics.

Colloid Osmotic Pressure. This force is due principally to the plasma albumin concentration (i.e., the albumin:water ratio). However, the quantitative relationships are interesting. The total osmotic pressure is approximately 200 times greater than the colloid osmotic pressure. The colloid osmotic pressure is 28 mm Hg, 19 mm of which are due to dissolved protein and 9 mm Hg of which are due to attracted Na^+ (the Donnan effect, see below). Globulins and other plasma proteins account for only 25% of the colloid osmotic pressure.

Since not all capillary pores are smaller than the diameter of plasma proteins, albumin leaks into the interstitial space. The total protein concentration here is 20 g/L, and this protein exerts a colloid osmotic pressure of close to 5 mm Hg (the total protein content of the interstitial space [20 g/L × 11 L] is almost equal to that in the vascular volume [80 g/L × 3 L]).

The interstitial fluid space is not uniform and exists in two phases, a liquid and a bound phase, the latter containing large anionic macromolecules (glucosaminoglycans).

The Donnan equilibrium is based on differences of concentration of impermeant ions (largely anionic albumin between the vascular and interstitial compartments). Albuminate attracts cations (largely Na^+) into and repels anions (largely Cl^-) out of the vascular compartment. Since the [Na] exceeds that of [Cl], the vascular space has a larger number of ionic species.

There is also a Donnan effect across the cell membrane due to the large intracellular macromolecular anions. In fact, there is a "double-Donnan," an equal force in the opposite direction due to the restriction of Na^+ to the ECF.

Suggested Reading

Halperin ML and Skorecki KL: Interpretation of the urine electrolytes and osmolality in the regulation of body fluid tonicity. Am J Nephrol 6:241–245, 1986.

Kregenow FM: Osmoregulatory salt transporting mechanisms: Control of cell volume in anisotonic media. Annu Rev Physiol 43:493–505, 1981.

Oh MS and Carroll HJ: Regulation of extra- and intracellular fluid composition and content. In Arieff AI and DeFronzo RA (eds): Fluid, Electrolyte and Acid Base Disorders. New York, Churchill Livingstone Inc, 1985, pp 1–38.

THE BIG PICTURE: Water Physiology

There is an obligatory loss of 0.8 L of water each day; therefore, mechanisms are needed to ensure this minimum water intake and to excrete any excess water.

Thirst: **Adequate water intake is ensured by the CNS thirst center. A rise in tonicity of only 1–2% provides a powerful urge to drink.**

Antidiuretic hormone (ADH): **ADH is the hormone that limits water excretion. A rise in tonicity or a fall in the "effective" blood volume causes ADH release from the posterior pituitary.**

Renal events: **for dilute urine to be excreted, ADH must be absent. The presence of ADH results in the excretion of a concentrated urine.**

Assess ADH action **on the kidney by 2 parameters—osmolality and volume: ADH produces a urine with** ***maximum*** **osmolality and usually a** ***minimum*** **volume. Conversely, in the absence of ADH, the urine has a** ***minimum*** **osmolality and** ***maximum*** **volume.**

Questions (Answers on page 151)

12. What can limit the renal excretion of "free water" in a subject drinking a large water load?

13. The plasma [Na] was 160 mM in a patient who was otherwise unaware of his problem. What aspect of his history should catch your attention?

14. Several days after a car accident, the patient has hypernatremia and mild ECF volume contraction. Three to 4 L of urine are passed/day (osmolality = 420 mOsm/kg H_2O). What is the most likely cause of polyuria?

15. How would you know that the failure to excrete "free water" was not due to an inadequate delivery of filtrate to the loop of Henle?

16. Two normal 70-kg persons excrete the following urine over the same time period:

Subject A, 1 L of urine (1200 mOsm/kg H_2O)

Subject B, 5 L of slightly hypertonic urine (450 mOsm/kg H_2O)

If neither has any fluid intake, who will have the higher plasma osmolality?

Answers

12. Fifty L of filtrate are delivered to the loop of Henle. Since the minimum urine osmolality is 50–60 mOsm/kg H_2O, 80% of this volume can yield free water (40 L). However, since some water reabsorption occurs in later nephron segments even in the absence of ADH (Table 6–2, page 155), less than half of this free water can be excreted. Thus, a normal kidney should be able to generate about 1 L of maximally dilute urine/hr.

To excrete 24 L of free water requires the excretion of 24 L × 50 mOsm/kg H_2O or 1200 mOsm/day (a normal diet yields 600 mOsm of urea and close to 200 mmol [400 mOsm] of NaCl and KCl daily). Therefore, in practice, patients taking in a very large water load (e.g., beer) with little protein or salt need to excrete endogenous urea and Na (Na loss is equivalent to a loss of ECF volume). Once ECF volume contraction occurs, this leads to ADH release despite hypoosmolality. In this circumstance, ADH limits further free-water excretion. Ultimately, the patient will suffer from severe hyponatremia (brain cell swelling) if water intake continues.

13. Hypertonicity should stimulate thirst! The absence of thirst in this patient should alert you to a CNS problem (hypothalamic lesion) involving the detection of thirst.

14. Although trauma and polyuria suggest diabetes insipidus due to a basilar skull fracture, this is unlikely because the urine osmolality is not a *minimum* value. The patient is under the influence of a diuretic. Since no drug was taken, either glucosuria or a urea load from tissue breakdown could cause an osmotic diuresis. The ECF volume contraction was due to Na loss in the urine during the osmotic diuresis.

15. If the urine contained Na plus K or Cl, NaCl was present at diluting sites.

16. Both subjects have 42 L of body water and an osmolality of 285 mOsm/kg H_2O before the urine losses (total mOsm are 11970 [285 × 42 L]).

Subject A has lost 1200 mOsm and 1 L; his new osmolality is 263 mOsm/kg H_2O ((11970 − 1200)/(42 − 1)L).

Subject B has lost 2250 mOsm (450 mOsm/kg H_2O × 5 L) and 5 L. His new osmolality is also 263 mOsm/kg H_2O ((11970 − 2250) mOsm/(42 − 5)L).

Therefore, both subjects have the same plasma osmolality. This calculation points out that one needs to know both the osmolality and the volume of urine in order to understand the contribution of renal excretion to body osmolality.

TONICITY DEFENSE INVOLVES THIRST AND FREE WATER EXCRETION OR CONSERVATION.

Maintenance of normal tonicity is virtually synonymous with regulation of total body water. Given appropriate access to water, control of body water (i.e., tonicity) is mediated via the hypothalamus, the posterior pituitary, and the kidney.

An increased tonicity should result in thirst and the prevention of renal free-water loss. Reduction of tonicity should diminish thirst and lead to free-water excretion. The controls of tonicity are remarkably sensitive, responding to a 1–2% change in "effective osmolality."

CONTROL OF WATER INTAKE

Thirst is stimulated by an increased tonicity (glucose and urea particles are not sensed by the hypothalamic centers). Contraction of the ECF volume is a weaker stimulus to thirst. Other factors unrelated to a need for water may also stimulate thirst (e.g., dryness of the mouth, habit, culture, psyche, etc.). The major inhibitors of thirst are hypotonicity and ECF volume expansion.

EXCRETION OF A DILUTE URINE (Fig. 6–4, page 153)

EXCRETION OF A DILUTE URINE REQUIRES:
- *DELIVERY* OF NaCl TO THE DILUTING SITES
- *REABSORPTION* OF SOLUTES WITHOUT WATER
- *EXCRETION* OF THIS FREE WATER

When excess water is ingested without solutes, water without solutes must be excreted in the urine. To excrete a dilute urine, three processes must occur:

Delivery of Saline to the Thick Ascending Limb of the Loop of Henle. Thirty-three percent of the GFR (60 L/day) is delivered to the diluting sites (Fig. 6–4A, page 153), i.e., 40 times more than is needed to excrete the usual daily water load of 1.5 L. Therefore, only a very large reduction in GFR can reduce delivery sufficiently to be the sole cause of a limited free-water excretion.

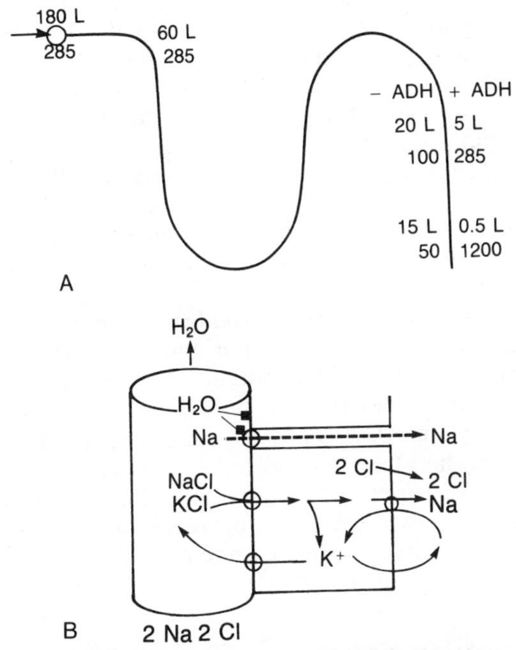

Figure 6–4. Luminal fluid volume and osmolality. *A*, The glomerulus is represented by the circle and the remainder of the nephron by the solid line. The volume of filtrate/24 hr is shown as the top number; the osmolality in the lumen at each nephron site is shown below (mOsm/kg H_2O). In the collecting duct, the numbers on the left represent the excretion of a dilute urine in the absence of ADH (−ADH); those on the right represent the excretion of a concentrated urine with ADH present (+ADH).

B, A portion of the thick ascending limb is shown in more detail in Figure 6–4B to illustrate the principal transporter units (a luminal membrane electroneutral Na, K, 2 Cl transporter and a basolateral NaKATPase; there are also a basolateral Cl conductance channel and both a luminal and a basolateral K conductance channel). Note also that the tight junction is permeable to Na^+ but not to water (nor is the luminal cell membrane permeable to water).

The principal sites at which ADH acts are at the insertion of the Na, K, 2 Cl transporters in the thick ascending limb, the induction of H_2O permeability in the cortical and medullary collecting duct, and the induction of urea permeability in the medullary collecting duct.

Separation of Salt and Water (Reabsorption of NaCl Without Water). This separation begins in the thick ascending limb of the loop of Henle; there is a net reabsorption of NaCl without water. This net NaCl transport is inhibited by furosemide which binds to the luminal Na^+, K^+, $2Cl^-$ transporter. To have a high luminal concentration, furosemide must be secreted in the proximal convoluted tubule; in renal insufficiency or with drugs inhibiting proximal secretion, this diuretic is less potent. In contrast, ethacrynic acid acts from the basolateral surface of this nephron segment.

In the collecting duct, the urine can be diluted further when NaCl is reabsorbed to a greater degree than water in the absence of ADH.

Maintenance of Separation. The hypotonic fluid that exits from the loop of Henle must not equilibrate with the iso- and hyperosmolar fluid surrounding the remainder of the nephron. These membranes remain relatively water-impermeable when ADH is absent; therefore, ADH secretion must cease when free water is to be excreted. ADH can be released for a variety of reasons unrelated to tonicity (see Chapter 7 on hyponatremia for more discussion of this point). Thus, a hyperosmolar urine can be excreted during a hypoosmolar state.

EXCRETION OF CONCENTRATED URINE (Figs. 6–4, page 153, and 6–5, page 155)

ADH IS REQUIRED TO EXCRETE A CONCENTRATED URINE.

When water intake is decreased, the tonicity of body fluids rises, owing to ongoing water loss. This stimulates the release of ADH from the posterior pituitary gland. Binding of ADH to its receptor on the renal collecting duct cell membrane results in the formation of cyclic AMP, which increases collecting duct permeability to water (Fig. 6–5, page 155). Therefore, water is reabsorbed passively down an osmotic gradient from the collecting duct lumen to the hyperosmolar renal medullary interstitium, resulting in the excretion of concentrated urine and the conservation of free water.

Hyperosmolar Medullary Interstitium. Both urea and Na contribute to the high medullary osmolality. The following

For the purpose of demonstration, consider the fate of 12 L of fluid delivered to the end of the loop of Henle.

TABLE 6–2. Effect of Water Reabsorption on Volume and Osmolality*

Nephron Site	Volume Delivered to (L)	Volume Reabsorbed by (L)	Osmolality (mOsm/kg H_2O)	Rise in Osmolality
End, Loop of Henle	12	—	100	—
End, Cortical Collecting Duct†	4	8	300	3×
End, Medullary Collecting Duct‡	1	3	1200	4×

*ADH makes the cortical and medullary collecting ducts permeable to water (Fig. 6–5). Note that the osmolality of the urine rises 3-fold in the cortex and 4-fold in the medulla; however, the bulk (67%) of the water is reabsorbed in the cortex, and this prevents "washing out" of solutes in the hypertonic medulla.

†67% water reabsorbed, no solutes.

‡75% water reabsorbed, no solutes.

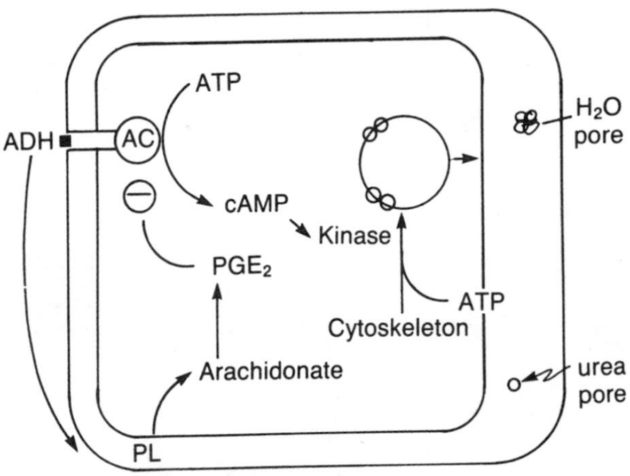

Figure 6–5. ADH action on the collecting duct. ADH binds to its cyclase-linked receptor on the basolateral membrane of collecting duct cells (left-hand side). This leads to activation of adenylate cyclase (AC) and to cyclic AMP formation and activation of a protein kinase. Resultant phosphorylations lead to the insertion of water channels or pores into the luminal membrane, making it permeable to water. Water moves passively to osmotic equilibrium. ADH also activates prostaglandin formation (PGE_2), which leads to a "turn-off" signal for water permeability. Hence, inhibitors of this latter system can prolong or potentiate ADH actions. ADH also leads to the insertion of pores for urea in the luminal membrane of the *medullary* collecting duct, promoting the increased medullary hyperosmolality. PL = phospholipid.

factors must act in concert to generate and maintain the hyperosmolar interstitium:

1. Active NaCl transport in the medullary thick ascending limb of the loop of Henle (adds solute to the interstitium).
2. Recirculation of urea in the medulla.
3. Preserved vascular architecture.
4. Slow medullary blood flow, so that solutes accumulated in the medulla are not "washed away."
5. ADH action contributes to the generation of this hyperosmolar interstitium.

- URINE OSMOLALITY = MEDULLARY FUNCTION + ADH ACTION
- URINE $[Na^+]$ + $[K^+]$ = RENAL IMPACT ON TONICITY

How Much of the Urine Osmolality Contributes to the Defense of Body Tonicity? To analyze the renal impact on tonicity, consider only the osmoles critical for water distribution across cell membranes and water itself, i.e., ignore urine urea (subtract the mM urine urea from the urine osmolality, as it does not influence tonicity). In contrast, the urine particles that are important for tonicity are Na and K (plus Cl and HCO_3).

Suggested Readings

Jamison RL and Oliver RE: Disorders of urinary concentration and dilution. Am J Med 72:308–322, 1982.

Robertson GL, Athar S, and Shelton RL: Osmotic control of vasopressin function. In Andreoli RE, Grantham JJ, and Rector FC Jr (eds): Disturbances of Body Fluid Osmolality. American Physiological Society, Baltimore, Williams & Wilkins, 1977, pp 125–148.

Schrier RW and Goldberg JP: The physiology of vasopressin release and the pathogenesis of impaired water excretion in adrenal, thyroid and edematous disorders. Yale J Biol Med 53:525, 1980.

PROCESSES INVOLVED IN THE GENERATION OF A HYPEROSMOLAR MEDULLARY INTERSTITIUM

Single Effect. To excrete a concentrated urine, there must be a hyperosmolar medullary interstitium (1200 mOsm/kg H_2O in humans). This seems to be the result of a single effect—the active transport of NaCl (via the Na, K, 2 Cl transporter) out of the lumen of the thick ascending limb of the loop of Henle. However, this can achieve only a 200 mOsm/kg H_2O gradient in any given horizontal plane. Longitudinal flow permits a progressive rise in absolute osmolality.

Selective Permeabilities. To generate a very hyperosmolar medullary interstitium, there are differential permeabilities of the descending and ascending limbs. The descending limb is water-permeable (and relatively solute-impermeable). Hence, luminal osmolality rises, owing to water loss as fluid descends in the loop. In contrast, the thin ascending limb is permeable to Na. Since the luminal [Na] rose in the descending limb owing to water abstraction, there is a favorable [Na] gradient for Na exit. The net result is electrolyte addition to the interstitium.

Role of Urea. In the cortex (late distal convoluted tubule and cortical collecting duct), water is reabsorbed under the influence of ADH. However, ADH does *not* induce urea permeability here. Thus the urea:H_2O ratio in the luminal fluid rises progressively (same urea but less water). In the medullary collecting duct lumen, there is now a much higher [urea] than in the medullary interstitium. Uniquely in this nephron segment, ADH induces *urea* permeability. Hence, urea diffuses passively into the medullary interstitium, adding solute to this critical area. This accounts for almost half of the hyperosmolality of the medulla.

Anatomy of Blood Vessels. The vasa recta descend straight to the medullary tip, then turn around and ascend in a parallel fashion. Given their permeability to water and solutes, they do not remove much material. They are called *countercurrent exchange vessels*.

In summary, the first three components generate medullary hyperosmolality and are called the *countercurrent multiplier*. The fourth component permits this hyperosmolality to remain. Low urea levels (low-protein diets), high vasa recta flow rates (renal vasodilatation), lack of ADH, or inhibition of the Na, K, 2 Cl transporter all compromise the ability to generate a hyperosmolar medullary interstitium. In addition, diseases involving this area also limit the ability to excrete a concentrated urine.

THE BIG PICTURE: Sodium Physiology

Na content determines the ECF volume; with ECF volume expansion, there is excess total body Na; with ECF volume contraction there is a Na deficit.

The renal response to ECF volume contraction is the excretion of a NaCl-free urine; failure to do so points toward a renal problem.

The major renal mechanisms preventing Na excretion when the ECF volume is low are a reduced GFR, enhanced proximal Na reabsorption, aldosterone-stimulated Na reabsorption (and K excretion), and increased medullary collecting duct Na reabsorption (? absence of atrial natriuretic factor (ANF)).

When a Na load is ingested, renal mechanisms are called into play to cause a natriuresis; these include an increased GFR, a relatively diminished proximal Na reabsorption, suppression of aldosterone release, and the release of ANF (Na is regulated within a 10% range).

The major conditions in which excessive Na is retained are heart failure and hypoalbuminemia.

Water is handled in an independent fashion from Na (requiring the hormone ADH (see Table 6–3, page 169).

Questions (Answers on page 159)

17. What is the total body Na content and the plasma [Na] in a patient with edema who retained an excessive quantity of water?

18. Identify the lesion in this patient: ECF volume contraction, plasma [Na] 130 mM and [K] 5 mM, urine [Na] 60 mM and [K] 10 mM. Is the plasma ADH high, low, or normal in this patient? (State your reasons for your answer.)

19. What are the volumes of distribution of the common intravenous solutions?

GENERAL FEATURES OF NA PHYSIOLOGY

The body contains close to 40 mmol Na/kg, and 94% is in the ECF. The major role of Na is to serve as an osmole

TABLE 6–3. Distribution Volumes of Commonly Used Intravenous Solutions*

Solution	Volume of Distribution (L)	
	ECF	*ICF*
Isotonic NaCl (150 mM, 0.9%)	1	0
D_5W	0.33	0.67
Half-normal NaCl (75 mM, 0.45%)	0.67	0.33
⅔ D_5W, ⅓ Isotonic NaCl	0.55	0.45
Hypertonic Saline (833 mM, 5%)	+3.9	−2.9
(500 mM, 3%)	+2.6	−1.6

*The volume of distribution of 1 L of commonly used intravenous solutions is shown for a 70 kg normal person.

Answers

17. The total body Na content is increased in any patient with edema. In this patient, excessive water is also retained, producing hyponatremia.

18. With ECF volume contraction, the urine should be Na-free. However, it is not! Similarly ECF volume contraction should cause aldosterone release and a high urine [K]. The low urine [K] suggests aldosterone deficiency as a cause for the renal Na loss.
The ECF volume contraction causes ADH release, which accounts for the failure to excrete free water.
Therefore, there are 2 reasons for hyponatremia, urinary Na loss and a diminished ability to excrete free water.
The underlying diagnosis in this patient is low aldosterone bioactivity.

19. The major premise is that isotonic saline remains entirely in the ECF, and free water distributes in proportion to the ICF and ECF volumes. Accordingly, the distribution of infusates can be calculated by dividing each solution into equivalent volumes of isotonic NaCl and "electrolyte-free water" (i.e., 1 L of 75 mM saline is 0.5 L of isotonic saline and 0.5 L of electrolyte-free water; see Table 6–3).

restricted to the ECF; Na and its anions account for 90% of the ECF osmoles. Since the circulating volume is part of the ECF, and since marked increases or decreases in circulating volume have serious consequences, there are important mechanisms to regulate total body Na content. The regulation of renal Na excretion is clearly the most important aspect of maintaining Na balance. In normal humans, about 17 mmol of Na are filtered/min (25,000 mmol/day), and more than 99% of this load is reabsorbed. It is important that reabsorption and filtration be linked, as small changes in GFR would otherwise produce large gains or losses in total body Na content. The phenomenon whereby changes in GFR are accompanied by parallel changes in tubular Na reabsorption is known as glomerulotubular balance.

Aldosterone has an important influence on renal Na reabsorption in the distal nephron; it is responsible for a small component (5%) of the total Na reabsorption. The secretion of aldosterone is stimulated by ECF volume contraction and suppressed by ECF expansion. In addition to GFR and aldosterone, other factors can influence renal Na reabsorption. In the proximal nephron, physical factors such as peritubular capillary colloid osmotic pressure and renal perfusion pressure modulate Na reabsorption. Atrial natriuretic factor (ANF) leads to a natriuresis when the ECF volume is expanded. ANF acts by increasing the GFR and decreasing medullary collecting duct Na reabsorption.

RENAL Na TRANSPORT

The renal handling of Na is best considered by examining the nephron segments individually (see also Table 6–4, page 161).

Na Handling in the Proximal Tubule. Fifty to 75% of filtered Na is reabsorbed in the proximal tubule, and electroneutrality is maintained by either Cl reabsorption or by H^+ secretion. The proximal tubular luminal membrane cannot maintain a large [Na] gradient, and this tubular epithelium is permeable to water; therefore, the osmolality of the fluid leaving the proximal tubule is the same as that of the ECF (Fig. 6–4, page 153). Na reabsorption down the [Na] gradient produced by the NaKATPase provides the driving force for nutrient and HCO_3 reabsorption by Na-dependent transporters (Table 6–5, page 163).

TABLE 6–4. Water and Na Reabsorption in the Kidney*

Nephron Site	Reabsorbed	Remaining	[Na]
Proximal Convoluted Tubule + Pars Recta	130 L 17500 mmol Na	180 − 130 = 50 L 7500 mmol Na	150
Loop of Henle and Early Distal Tubule	5L 5500 mmol Na	50 − 5 = 45 L 2000 mmol Na	45
ADH Present			
Cortical Collecting Duct	40 L 1250 mmol Na	45 − 40 = 5 L 750 mmol Na	150
Medullary Collecting Duct	4 L 600 mmol Na	5 − 4 = *1 L* *150 mmol Na*	150
No ADH Present‡			
Cortical Collecting Duct	15 L 1250 mmol Na	45 − 15 = 30 L† 750 mmol Na	25
Medullary Collecting Duct	10 L 600 mmol Na	30 − 10 = *20 L‡* *150 mmol Na*	*7.5*

*Renal events: Assuming that an individual has a normal GFR of 180 L/day, the following depicts the events involved in the excretion of 150 mmol of Na.

†In the collecting duct, Na is reabsorbed to a greater extent than water. Hence, dilution of the urine also occurs in this nephron site. In quantitative terms, 40% of the fall in urine osmolality occurred here in this example.

‡In the total absence of ADH, the collecting duct epithelium still has some water permeability. Thus, some water moves along the osmotic gradient into the interstitium. Note that more water is reabsorbed in the medulla in the absence than in the presence of ADH (owing to the greater water delivery)!

Na Handling in the Loop of Henle. NaCl reabsorption in the loop of Henle enables the excretion of a concentrated or dilute urine (for details, see Figure 6–4, page 153). NaCl is actively transported in the thick ascending limb, utilizing the luminal electroneutral Na^+, K^+, 2 Cl^- transporter, and, again, the driving force is provided by the basolateral NaKATPase, but only half of the Na ions use the intracellular route. Absolute Na reabsorption in the loop of Henle varies directly with delivery. Net NaCl reabsorption is stimulated by ADH in the medullary thick ascending limb, and NaCl reabsorption is inhibited by loop diuretics.

Na Handling in the Distal Convoluted Tubule. The major function of this segment of the nephron is Na balance and free-water formation. Na is actively reasorbed in this nephron segment along with Cl; high transepithelial [Na] gradients can be maintained.

Na Handling in the Collecting Duct. The major functions of the cortical collecting duct are K and H^+ secretion and Na reabsorption. In the medulla, the major functions are urine concentration, NH_4 excretion, and Na balance. There is active Na transport in the collecting duct and mineralocorticoids play an important role in this transport in the cortex. Na reabsorption may be accompanied by Cl reabsorption or by the secretion of K^+ and H^+. The collecting duct is also capable of maintaining large [Na] gradients across its epithelium. Active H^+ secretion occurs in the cortical and medullary collecting duct. The high transluminal potential difference in the cortical collecting duct that is due to Na^+ reabsorption can augment K^+ and H^+ secretion at this site.

REGULATION OF Na EXCRETION

One should expect to see an almost Na- (and Cl-) free urine when the ECF volume is contracted and to see Na excretion when the ECF volume is expanded. During euvolemia, the kidney excretes the dietary NaCl load. Hence, there are *no normal values* for urine Na and Cl; these must be interpreted relative to the physiologic state of the patient.

Na (and K) excretion can be influenced by the anion composition of the filtrate. For example, a patient who has vomited delivers more HCO_3^- to the collecting duct than can be reabsorbed. In this example, the urine contains Na^+ (and K^+) in conjunction with HCO_3^-, which is not reabsorbed despite the fact that the ECF volume is contracted; however, the urine does not contain Cl^-.

TABLE 6–5. Summary of Transporters in the Nephron

Nephron Site	Luminal Transporter	Function	Inhibitor
Proximal convoluted tubule	• Na/H^+ • Na-nutrient • NaCl	• Na + HCO_3 recycled • Nutrient reabsorbed • ECF volume regulation	• CAI* • Phloridzin
Loop of Henle	• Na/K/2 Cl	• NaCl reabsorbed • Dilution • Hyperosmolar medulla	• Furosemide
Distal convoluted tubule	• NaCl	• NaCl reabsorption • Dilution	• Thiazides
Cortical collecting duct	• Na^+ channel	• K^+ secretion • Na^+ reabsorption	• Amiloride • Triamterene • Spironolactone
	• H^+ ATPase	• NH_4^+ excretion	
Medullary collecting duct	•? NaCl • Na^+ channel	• Na balance • Na balance	•? ANF • Amiloride • Triamterene
	• H^+ ATPase	• NH_4^+ excretion	

*CAI = carbonic anhydrase inhibitors, which do not block the Na/H antiporter but inhibit HCO_3 reclamation (see Chapter 1 for more details).

TABLE 6–6. Effects of Aldosterone Deficiency on Urine Dilution

ADH-Independent
Diminishes Na delivery to diluting sites by low GFR
Thick ascending limb
Collecting duct
Diminishes solute reabsorption in the cortical collecting duct
ADH-Dependent
Low "effective" circulatory volume causes ADH release for nonosmotic reasons

TABLE 6–7. Effect of Glucocorticoid Deficiency on Urine Dilution

ADH-Independent
Diminishes hemodynamics, thereby reducing Na delivery to diluting sites (not a direct effect on ADH-independent water permeability)
ADH-Dependent
Elevates and sustains ADH-induced cyclic AMP rise
Low "effective" cardiac performance causes ADH release for nonosmotic reasons

Questions (Answers on page 165)

20. What are the purposes for administration of the various IV solutions?

21. When should hypertonic NaCl be infused? How much should be given?

Suggested Reading

de Bold AJ: Atrial natriuretic factor: a hormone produced by the heart. Science 230:767–770, 1985.

Marsden PA and Skorecki KL: Afferent limb of volume homeostasis. In Brenner BM and Stein JH (eds): Continuing Issues in Nephrology. Volume 16. New York, Churchill Livingstone, Inc., 1987, pp. 1–32.

Rector FC Jr: Sodium, bicarbonate, and chloride absorption by the proximal tubule. Am J Physiol 244:F461–F471, 1983.

Steinmetz PR and Koeppen BM: Cellular mechanism of action of diuretics along the nephron. Hosp Pract 19:125–134, 1984.

Answers

20. There are four different impacts that intravenous infusions have on body fluid compartments. They do the following:

Expand the ECF Only. Isotonic saline is the solution to infuse in patients with ECF volume contraction. If the vascular volume is especially low, add albumin, plasma or blood.

Expand the ICF With as Little Fluid to the ECF as Possible. Infuse water with a particle that prevents osmotic lysis of blood cells but one that disappears promptly. The water disperses, ⅔ in the ICF and ⅓ in the ECF. Glucose is the most commonly used particle, but *beware*—not all patients can metabolize large quantities of glucose rapidly. This is used in patients with ICF volume contraction (hypernatremia).

Expand Both the ICF and ECF Volumes. With salt and water deficits, a dilute saline solution is infused. The first option is half-normal saline (75 mM, 0.45%), which distributes ⅔ in the ECF; the second option is ⅔ D_5W and ⅓ NaCl, which distributes close to 50% in each compartment. This latter solution requires glucose removal.

Remove Water From Cells. Hypertonic saline (3%, 5%) causes a water shift from cells. However, a necessary concomitant event is expansion of the ECF volume. This is used in a patient with *symptomatic* hyponatremia.

21. Hypertonic saline is used to draw water out of cells, especially brain cells. The driving force for this water shift is a rise of tonicity in the ECF (a rise in the Na:H_2O ratio). This therapy has the major disadvantage of expanding the ECF volume (see page 185 of Chapter 7).

7
HYPONATREMIA

THE BIG PICTURE

General Considerations

- ***Definition:*** **[Na] Below lower limit of normal (136 mM)**
- **Rule out pseudohyponatremia, a laboratory error due to hyperlipidemia or hyperproteinemia. Rule out hyperglycemia**
- ***Causes:*** **Primary Na loss and/or primary water gain**

Cell Volume Changes. **With the exceptions of pseudohyponatremia and hyperglycemia, hyponatremia leads to a water shift into cells and cell swelling. This is especially important in the CNS because the brain is enclosed in a fixed space. Treatment must be aggressive if symptoms are present.**

Hyponatremia Due to Na Loss. **Na loss leads to ECF volume contraction. To determine the cause of Na loss, examine the urine electrolytes: The urine should have a minimal [Na] or [Cl]; otherwise, look for a renal problem. Treatment consists of Na replacement and attention to underlying problems.**

Hyponatremia Due to Water Gain. **Water gain is**

PHYSIOLOGY CAPSULE

[Na] = ICF Volume

Remember the $[Na^+]$ is the $Na^+:H_2O$ ratio! The ICF volume can be deduced from the [Na]; it is contracted in hypernatremia and expanded in hyponatremia. The only exceptions to this rule are pseudohyponatremia and the addition of a "Na-like" particle to the ECF, e.g., glucose or mannitol (a "Na-like" particle is one that is largely restricted to the ECF).

Na Content = ECF Volume

Na is an ECF osmole and is in the body to keep water out of cells; thus, *Na content* reflects the ECF volume. Na deficit means ECF volume contraction, and Na excess means ECF volume expansion.

Danger of Hyponatremia

In those cases of hyponatremia with an expanded ICF volume, the danger is brain cell swelling. On a chronic basis, brain cells tend to regulate their volume largely by ion extrusions.

Renal Response to Na Loss

The appropriate renal response to a Na deficit (ECF volume contraction) is to avoid additional loss of Na or water in the urine. Check for these responses: the urine should have a very low [Na] and [Cl], and the urine osmolality should be high (a reduced "effective" vascular volume leads to ADH release).

Renal Response to Water Excess

The appropriate renal response to water excess is to excrete the *maximum* volume of *maximally* dilute urine (50 mOsm/kg H_2O). Should this not be observed, suspect that ADH is present in a significant concentration in plasma (seek the cause!) or that the kidneys are abnormal.

almost always due to limited renal water excretion. A normal kidney can make 12 L of "free water"/day; thus, excessive water intake alone is virtually never the primary cause of hyponatremia. Reduced "free-water" excretion is usually due to ADH action; ADH is released from the posterior pituitary when there is a reduced "effective circulatory volume," hypertonicity, neural stimuli, or drug ingestion.

***Therapy*. Deal with the underlying disorder, limit water intake, and promote "free-water" loss. If there are CNS symptoms, give hypertonic saline.**

ILLUSTRATIVE CASE

A healthy young male developed aseptic meningitis 5 days ago. This was accompanied by frequent vomiting. He took no medications. Physical examination revealed ECF volume contraction only on day 3. Laboratory results are summarized in the table below.

	Day 3	Day 5
Plasma		
Na (mM)	130	117
K (mM)	3.1	4
HCO_3 (mM)	30	24
Osmolality (mOsm/kg H_2O)	270	245
Glucose mg/dl (mM)	90 (5)	90
Creatinine mg/dl (μM)	2.3 (200)	1.1 (100)
Urine		
Na (mM)	60	50
Osmolality (mOsm/kg H_2O)	450	407

Questions (Answers on page 169)

1. Could Na loss have caused the hyponatremia on day 3? Is the ADH level elevated in this patient? (State the reasons for your answer.)
2. Why is the plasma [Na] lower on day 5? Is the ADH level higher?
3. What is the best urine electrolyte to indicate ECF volume contraction? Would your answer change if the patient had diarrhea rather than vomiting?

APPROACH TO HYPONATREMIA

The approach to hyponatremia is summarized in Flow Chart 7–A, page 171. Pseudohyponatremia should first be ruled

TABLE 7–1. Causes of Hyponatremia

Pseudohyponatremia (Normal Plasma Osmolality)
- Hyperlipidemia, hyperproteinemia

Addition of Particles Largely Restricted to the ECF
- Hyperglycemia, mannitol

Decreased "Effective Circulating" Volume
- ECF volume contraction (loss of Na^+]
- ECF volume *NOT* low (edema states)
  - Low vascular volume
    - Hypoalbuminemia (liver, kidney, or GI problem)
    - Leaky capillaries to albumin (e.g., sepsis)
  - Low arterial volume but expanded venous volume
    - Heart problems

Primary Water Gain (and Secondary Na Loss)
- ADH release from the posterior pituitary without a tonicity or "effective circulatory volume" stimulus
  - Drugs (see Table 7–4, page 183), excessive pain, emesis, vagal nerve stimulation, CNS disease, metabolic disorders (e.g., porphyria)
- ADH from other sources (neoplasms, exogenous administration)
- Drugs potentiating ADH action (see Table 7–4)
- Drugs interfering with free-water formation (e.g., loop diuretics)

Answers

1. Yes. Not only does Na loss directly cause hyponatremia, it also leads to ECF volume contraction (note elevated plasma creatinine, low plasma [K], and elevated plasma $[HCO_3]$). A low ECF volume causes ADH release and thereby prevents free-water loss in the urine. Patients with meningitis can have ADH release independent of their tonicity and ECF volume. We know that the ADH level is elevated because the urine osmolality is high. In addition, continuing water intake in this situation would lead to water retention and more severe hyponatremia.

2. The lower serum [Na] occurred because the patient lost more Na and/or had a large water intake. The absolute ADH level cannot be deduced from the plasma [Na], but it is high enough to prevent renal free-water excretion.

3. Following NaCl loss, the urine should have minimal Na and Cl (<10 mM) if kidney function is normal. In this case, vomiting results in the loss of ECF Cl and a commensurate rise in HCO_3^-. The HCO_3^- filtered load rises, and this excess cannot be reabsorbed. The resultant Na^+ loss leads to ECF volume contraction, renin release, angiotensin II formation, and, thereby, aldosterone release. Aldosterone causes K^+ loss in the urine. Hence a very low urine [Cl] but not [Na] is expected (Fig. 4–1, page 99).

$NaHCO_3$ loss with diarrhea produces ECF volume contraction and a low plasma $[HCO_3]$. The kidney responds to acidemia by excreting a large quantity of NH_4^+ (with Cl^-). Thus, in this case there is a very low urine [Na] but not a very low urine [Cl].

In summary, in the first case, the urine [Cl] is a more reliable index of ECF volume contraction, whereas in the second case the urine [Na] is more reliable.

out; this can be done by examining for high lipids (turbid serum) or high proteins, as in myeloma. In the presence of pseudohyponatremia, the serum osmolality is not low.

> 1. RULE OUT PSEUDOHYPONATREMIA AND THE HYPONATREMIA SECONDARY TO HYPERGLYCEMIA (OSMOLALITY NOT LOW).
> 2. IF HYPONATREMIA IS ASSOCIATED WITH ECF VOLUME CONTRACTION, Na HAS BEEN LOST.
> 3. IF HYPONATREMIA IS ASSOCIATED WITH A NORMAL OR HIGH ECF VOLUME, LOOK AT THE "EFFECTIVE" CIRCULATING VOLUME—LOW CARDIAC OUTPUT SUGGESTS A CIRCULATORY PROBLEM, AND NORMAL VALUES SUGGEST AUTONOMOUS ADH RELEASE.

Hyperglycemia should also be ruled out; again, the plasma osmolality is *not* low in this type of hyponatremia. However, not all cells are affected equally by hyperglycemia (see Chapter 12, Fig. 12–1, page 279).

All causes of hyponatremia with a low osmolality include an expanded ICF volume. The brain is most affected because it is in a "closed box." Therapy is designed to reverse brain swelling. The price to pay for this therapy is expansion of the ECF volume (see page 185).

The primary cause of hyponatremia is Na loss or water gain. The latter is due to water intake plus ADH action. If a low "effective" circulating volume is not present, autonomous ADH release is the diagnosis, and the causes are discussed in the SISADH section of this chapter, page 178.

CAUSES OF HYPONATREMIA (Table 7–1, page 169)

Pseudohyponatremia

> - IN PSEUDOHYPONATREMIA (HYPERLIPIDEMIA, LIPEMIC SERUM), THERE IS A NORMAL SERUM OSMOLALITY
> - CELLS ARE NOT SWOLLEN IN PSEUDOHYPONATREMIA

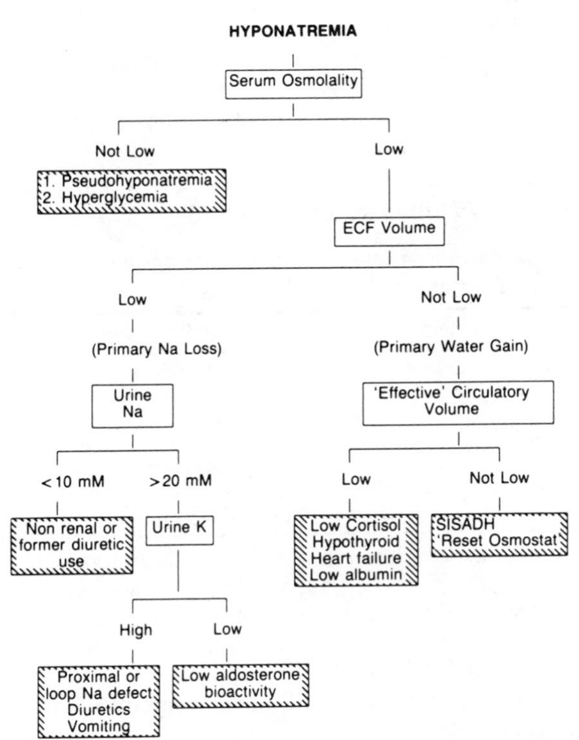

Flow Chart 7–A. Approach to hyponatremia. The clear boxes represent the variables to be evaluated clinically to reach a final diagnosis. The final diagnoses are shown in the shaded boxes (see text for details).

Na is dissolved only in the water phase of plasma. The presence of excessive quantities of *nonaqueous* volume in the plasma causes a fall in the Na:total volume ratio. Recall that the [Na] of plasma is 140 mM (i.e., per L total volume), whereas the true [Na] (Na:H_2O ratio) is 152 mmol/L of *water*. Thus, the discrepancy is apparent rather than real because the Na content is the same; it is just distributed in a different volume (plasma is only 93% water). Normally, lipid and protein make up the nonaqueous volume of plasma, and either of these may increase. This diagnosis can be confirmed by inspecting plasma for lipid or measuring the plasma osmolality—a normal value is found in pseudohyponatremia. If a Na^+-selective electrode were used on undiluted plasma, the [Na] (Na:H_2O ratio) would approach 152 mmol/kg H_2O.

Questions (Answers on page 173)

4. Match the 4 separate causes of hyponatremia with the critical parameter for diagnosis.

Etiology	Critical Parameter Measured
(1) Diuretic intake	a) Normal plasma osmolality
(2) Hyperlipidemia	b) Normal ECF volume and low urine osmolality
(3) Excess ADH	c) ECF volume contraction and hypokalemia
(4) Compulsive water drinking	d) Low serum and high urine osmolality

5. A diabetic has renal failure. His blood sugar is 1000 mg/dl (55 mM), [Na] is 128 mM, and blood urea nitrogen is 70 mg/dl (urea 25 mM). What is his calculated osmolality, and has hyperglycemia changed his ECF and ICF volumes?

6. What is the average 24-hour urine osmolality in a normal subject?

7. What conclusions can be drawn when the urine osmolality is high?

8. Three patients have hyponatremia (120 mM): in Case A it is due to water gain; in case B, Na loss; in Case C, hyperglycemia. Which patient(s) will have brain cell swelling? (Give reasons for your answer.)

Hyponatremia Due to Hyperglycemia

In contrast to pseudohyponatremia, hyponatremia due to

Answers

4. (1) = c, (2) = a, (3) = d, (4) = b.

5. Hyperglycemia causes water to shift out of muscle cells (the bulk of the ICF volume). Quantitatively, a rise in [glucose] of 900 mg/dl causes the [Na] to fall by 12 mM (9×1.35) to 128 mM (see page 174 for details concerning this calculation). The calculated osmolality is $2 \times$ [Na] + [glucose] + urea in mM units or 336 mOsm/kg H_2O. The ECF volume is expanded because the muscle ICF volume has contracted.

The liver ICF volume is expanded and the brain volume is probably close to normal (see Chapter 12, Fig. 12–1, page 279).

6. A normal person produces 500 mmol of urea and consumes approximately 200 mmol of NaCl plus KCl; thus, the urine contains 900 mOsm (NaCl + KCl = 400 mOsm). If a typical urine volume is 1.5 L, the osmolality of a 24-hour urine is close to 700 mOsm/kg H_2O.

7. A high urine osmolality indicates that ADH is acting on the kidney and that the renal medulla has a high osmolality. However, this concentrated urine may lead to hypernatremia or hyponatremia. Consider the following two examples in which nothing was ingested. First, the excretion of 1 L of hyperosmolar urine (600 mOsm/kg H_2O), when urea is the sole osmole, leads to a rise in the ECF [Na] (the Na:H_2O ratio rises) because water without Na was excreted. In contrast, the excretion of a urine with the same osmolality with a [Na] of 200 mM (higher than the normal Na:H_2O) favors the development of hyponatremia.

8. *Cases A & B:* The brain cells are swollen in both of these patients and to the same degree as the [Na] is equal. The [Na] tells you the ICF volume providing no additional particles (like glucose in Case C) are acting to shift water out of cells, and defense of the ICF volume has not occurred.

Case C: Hyponatremia due to hyperglycemia causes muscle cells to shrink, liver cells to swell, and brain cells to remain close to normal in size (see Chapter 12, Fig. 12–1, page 279).

hyperglycemia is real. In this case, the water content of the ECF rises because glucose draws water out of muscle cells by osmosis (see Chapter 12, Fig 12–1, page 279). In quantitative terms, expect a 1.35 mM [Na] fall for every 100 mg/dl (5.5 mM) rise in the glucose concentration or the [Na] falls 1 mM with a 4 mM rise in blood glucose concentration. This calculation incorporates the fact that the glucose concentration in the ICF of the liver (and other noninsulin-sensitive cells with respect to glucose transport) is the same as that in the ECF. In this setting, the osmolality is not low, which it is in all other causes of true hyponatremia.

Hyponatremia With Hypoosmolality (Table 7–1, page 169)

> THE PHYSIOLOGIC (OR APPROPRIATE) STIMULI FOR ADH RELEASE ARE HYPERTONICITY AND DECREASED "EFFECTIVE" CIRCULATING VOLUME.

To understand the cause of hyponatremia, you should look at the input and output of Na and water. On the input side, a typical intake of Na is close to 150 mmol per day. The daily water intake is usually 1.5 L. On the output side, to stay in balance, a normal subject would have to excrete 150 mmol of Na in 1.5 L of urine; thus, the 24-hour urine [Na] should be close to 100 mM. Therefore, with a typical diet, hyponatremia is produced if the urine [Na] exceeds 100 mM. Note that the patient's diet and urine [Na] and *NOT* the osmolality are critical factors in understanding the renal contribution to the cause of hyponatremia (see Answer 7 on page 173).

> - MEASUREMENT OF THE URINE OSMOLALITY HELPS ASSESS ADH ACTION AND MEDULLARY HYPERTONICITY.
> - MEASUREMENT OF THE URINE [Na] HELPS TO DETERMINE THE CAUSE OF HYPONATREMIA.

Hyponatremia Caused by a Low "Effective" Volume (Table 7–2, page 175)

TABLE 7–2. Cause of "Effective" Circulatory Volume Contraction

ECF Volume Depletion (Na Loss)
Nonrenal Na Loss
Gastrointestinal tract
Vomiting, drainage, ileus, diarrhea
Skin
Excessive sweating, burns
Renal Na Loss
Diuretics
Drugs
Osmotic (glucose, urea)
Low aldosterone
Tubular disorders
Proximal (Fanconi syndrome, etc.)
Loop (Bartter's syndrome?)
Distal (interstitial disease, etc.; low aldosterone)
ECF Volume Normal but Maldistributed
High Interstitial and Low Vascular Volume
Low plasma albumin
(e.g., liver disease, nephrotic syndrome)
Albumin leaks out of capillaries
Low Arterial Volume, High Venous Volume
Primary myocardial, valvular, or pericardial disease

- Na MAINTAINS THE ECF VOLUME
- LOSS OF Na (ECF VOLUME) STIMULATES ADH RELEASE
- ADH PREVENTS THE EXCRETION OF FREE WATER

ECF Volume Contraction. Hyponatremia is almost always due to reduced water excretion in the face of a hypotonic intake. Loss of Na without water also causes hyponatremia. The mechanism for the Na-loss type of hyponatremia usually involves 2 steps: first, the loss of Na lowers the numerator of the Na:H_2O ratio. Second, and more indirectly, loss of Na results in ECF volume contraction and, as a result, ADH release from the posterior pituitary. ADH prevents the excretion of dilute urine and can produce hyponatremia when water is ingested. Thus, it is misleading to call this a "depletional hyponatremia," since there is usually a "dilutional" component as well.

ECF Volume Not Contracted. A patient need not have a reduced ECF volume to have ADH release on a "volume basis." Even if the total ECF volume is normal or high, the vascular component may not be "pumped in an optimum fashion" by the heart. The term "effective" circulating volume is used to refer to the critical component of the ECF required for adequate tissue perfusion (notice the vague terms in this description). In a more practical clinical sense, this "effective" volume is decreased either when the overall ECF volume is reduced or when the ECF volume is distributed such that there is insufficient volume in the vascular space. This latter maldistribution occurs in some edema states (e.g., hypoalbuminemia). A second type of maldistribution occurs when the arterial volume is low and the venous volume is high (when there is a primary decrease in cardiac function; see Table 7–2, page 175).

Questions (Answers on page 177):

9. Could Na loss have occurred via the renal route if the urine is now Na-free?

10. What role might K depletion play in determining the severity of hyponatremia?

11. How much water must a person drink to produce hyponatremia? Would it matter if the subject were taking a low-salt, low-protein diet?

Answers

9. Yes. If the patient took the diuretic yesterday and had the natriuresis then (and did not ingest Na today), he would have ECF volume contraction and the appropriate renal response of minimal Na and Cl in the urine today. This example was presented to emphasize that you must suspect the antecedent intake of diuretics (which might even be denied by the patient).

10. K depletion is expected when there is hyponatremia and decreased ECF volume because the ECF volume contraction leads to aldosterone release, which augments K excretion. Where does this urinary K come from? As shown in Chapter 9, Figure 9–7, page 227, much of this K was derived from muscle ICF and required Na to shift from the ECF to the ICF. This depletes ECF Na and further accentuates the degree of hyponatremia (this portion of hyponatremia is corrected by KCl therapy).

11. A person with a normal kidney can excrete at least 12 L of free water a day. Hence, water intake that exceeds this amount can be the sole cause of hyponatremia; this is called psychogenic polydipsia. Recall that each L of maximally dilute urine contains at least 50 mOsm. To excrete 15 L, the subject needs 15 L $\times$ 50 mOsm/L, or 750 mOsm. Failure to provide these in the diet results in the excretion of endogenous osmoles. When urea "runs out," the patient should start to lose electrolytes. Loss of Na results in ECF volume contraction and, thereby, ADH release. This obviously compromises further free-water excretion and can lead to severe hyponatremia. This pattern can be seen in patients with an excessive beer intake. Thus, bartenders who serve excretable osmoles (salty nuts and pickled eggs, which contain protein that can be converted to urea) at the bar are "practicing physiologists," because their patrons can now drink large volumes of beer without suffering the dangers of severe hyponatremia (beer contains only a few mmol of Na, K, and Cl but close to 1000 mmol of ethanol).

Adrenal Insufficiency. Hyponatremia is commonly seen in patients with adrenal insufficiency. The major mechanisms are renal Na wasting and reduced water excretion secondary to ADH release due to ECF volume contraction (a low aldosterone consequence) and reduced cardiac performance (a low glucocorticoid consequence). It is no longer believed that glucocorticoids influence water permeability in the collecting duct. Hyponatremia due to adrenal insufficiency is obvious from clinical grounds and from studies on renal K handling and hyperkalemia. *Caution:* These patients can be in great danger if given a water load or a K load or if they have another cause for Na loss (e.g., diarrhea). Treatment with replacement doses of glucocorticoids and mineralocorticoids and replacing the Na deficit corrects the hyponatremia.

A decreased "effective" circulating volume provides a central stimulus to ADH release, primarily via low pressure (venous) receptors, even in the presence of hypotonicity. It may also stimulate thirst (water intake must be present to develop hyponatremia).

Causes of Na Loss. Na can be lost by nonrenal and renal mechanisms. The nonrenal sources are usually obvious (gastrointestinal tract, skin) and are suspected by finding a urine that is virtually Na- and Cl-free. The urine [K] and osmolality may be high despite hyponatremia.

The most common renal cause for Na wasting is the ingestion of a diuretic; less often, an osmotic diuretic (glucose, urea) may cause Na loss. There are also a number of kidney diseases associated with renal Na wasting. The most useful index to discriminate among them is to examine K excretion. Low urine [K] in the face of renal Na loss and ECF volume contraction should suggest low aldosterone bioactivity (see page 179 and Chapter 10 for details). In contrast, a high urine [K] in the renal Na wasters suggests that the problem is in the proximal convoluted tubule, loop of Henle, or early distal convoluted tubule.

Hyponatremia Due to ADH Release That is not Caused by a Decrease in the "Effective" Circulating Volume (Table 7–3, page 179)

ADH RELEASE IN SISADH IS FROM THE POSTERIOR PITUITARY OR FROM A NONPHYSIOLOGIC SOURCE (e.g., A NEOPLASM).

ADH release in the absence of tonicity, "effective volume," and other physiologic stimuli is said to be "inappropriate." Thus, patients with this type of a high ADH state are said to have a *"syndrome of inappropriate secretion of ADH"* or SISADH for short. The causes for this high ADH level can be divided into two major categories: ADH release from the normal gland (posterior pituitary) and ADH or ADH-like material from other sources. A list of causes of high ADH levels is presented in Table 7–3, page 179.

What are the consequences of a persistently elevated ADH level? In general, the clinical effect is to prevent the excretion of free water. Thus, if water is not ingested or administered, the patient will suffer no consequences from

TABLE 7–3. Origin for the High ADH in Hyponatremia

ADH From Posterior Pituitary Gland
Cardiovascular (low "effective" circulating volume)
Specific diseases
Pulmonary lesion
CNS lesion
Endocrine disorders (e.g., hypothyroidism, hypoadrenalism)
Metabolic disorders (e.g., acute intermittent porphyria)
Excessive pain (e.g., during postoperative period) or vomiting
Drug-induced (see Table 7–4, page 183)
Mechanism unclear
ADH From Other Sources
Solid neoplasms (especially oat-cell carcinoma of lung)
Possibly granulomas such as tuberculosis
Exogenous administration
ADH (e.g., treatment for diabetes insipidus)
Oxytocin for labor induction
Potentiation of Endogenous ADH by Drugs (Table 7–4)
Adenylate cyclase activation
Phosphodiesterase inhibition
Adenosine receptor antagonists
Prostaglandin inhibitors*

*Prostaglandins antagonize ADH action by inhibiting cyclic AMP formation. Prostaglandin synthesis is also stimulated by ADH. This diminishes the ADH-induced cyclic AMP rise. In addition, the vasodilator action of prostaglandins increases medullary flow, which lowers medullary hyperosmolality. Both of these actions aid in the excretion of a dilute urine.

a higher ADH level. When water is ingested, it is retained, leading to hyponatremia with initial ECF volume expansion. In response to this ECF volume expansion, Na is excreted in the urine. Hence, the characteristic findings are hypotonicity, hyponatremia, and the absence of ECF volume contraction, together with the excretion of urine with an osmolality greater than the minimum value of 50 mOsm/kg H_2O and a urine that contains an appreciable amount of Na.

In summary, before confirming the diagnosis of SISADH, be sure that there is no reduction of the "effective" circulating volume. You can subclassify patients with SISADH into 4 subcategories (Fig. 7–1, page 181). Finally, a patient with SISADH may also have ECF volume contraction for other reasons. In this case, the urine [Na] is very low. However, hyponatremia is not corrected after the Na deficit is replaced.

Auxiliary Tests in Patients With SISADH

Other laboratory values that are helpful in the differential diagnosis of SISADH are as follows:

[K]. In the SISADH syndrome, the plasma [K] is usually normal; therefore, abnormalities in the plasma [K] should alert you to seek other causes for hyponatremia. When the ECF volume is low, aldosterone is released, which results in renal K loss. This loss is especially high when "cortical distal nephron" volume delivery is high. Therefore, diuretic- or vomiting-induced hyponatremia should be suspected if hypokalemia is present. In contrast to the above, suspect low aldosterone states if hyponatremia is accompanied by hyperkalemia.

[HCO_3^-]. In SISADH, the plasma [HCO_3^-] is usually normal, but it may be high in diuretic-induced hyponatremia (see Chapter 4). In contrast, hyponatremia of hypoaldosteronism is generally accompanied by a mild fall in the plasma [HCO_3^-] (to close to 20 mM).

Blood Urea. In patients with SISADH, urea clearance increases. This, together with dilution due to water retention, results in a fall in the BUN. In contrast, hyponatremia due to a low "effective" circulating volume as the stimulus to ADH release leads to a fall in the GFR and thereby to a rise in the BUN.

Urate. The urate level may also be reduced in patients with SISADH.

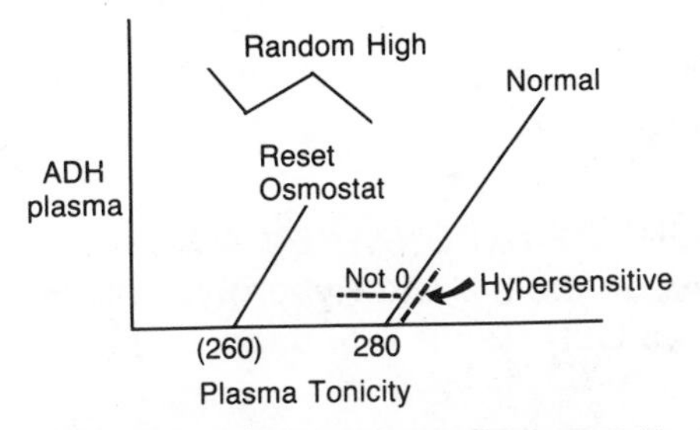

Figure 7–1. Subclassification of patients with SISADH (drawn from the original by GL Robinson). There are 4 subgroups of SISADH: (1) Random high autonomous ADH release (e.g., carcinoma of the lung). They represent 38% of SISADH. (2) Reset osmostat. They have normal regulation of ADH release but it is focused around a hypotonic stimulus. About 33% of patients with SISADH fall into this subcategory. (3) Failure to suppress ADH totally with hypotonicity. About 16% of patients have this disorder, in which ADH release is normal at high tonicity but is not 0 at low tonicity. (4) The fourth subgroup (14%) of SISADH have no problem with ADH, but their kidneys are either overly sensitive to it or there is a non-ADH–ADH-like material present.

CLINICAL DIAGNOSIS

Hyponatremia is a very common fluid and electrolyte disorder. Although you can suspect that it will develop in certain clinical situations associated with a decreased "effective" circulating volume and water administration, its presence is obvious when the lab results are examined. The clinical approach should answer the following questions:

1. Is there a decreased "effective" circulating volume?
2. Are there drugs that promote ADH release or action?
3. Is there a lung or CNS lesion that is responsible for ADH release?
4. Does the patient respond appropriately to a small water load?

If the answers to questions 1 and 2 are "no," then the patient has SISADH.

TREATMENT OF HYPONATREMIA (After Pseudohyponatremia and Hyperglycemia Have Been Ruled Out) (Table 7–5, page 183)

> 1. HYPONATREMIA IS NOT A SPECIFIC DISEASE—IT IS A FLUID AND ELECTROLYTE "SYMPTOM."
> 2. TWO SPECIFIC LINES OF ATTACK SHOULD BE CONSIDERED:
>   - DEAL WITH THE UNDERLYING DISEASE (e.g., IMPROVE THE ARTERIAL CIRCULATION IF NECESSARY)
>   - WITHHOLD WATER AND PROMOTE WATER LOSS
>   - RAISE THE [Na] TOWARDS NORMAL. SPEED DEPENDS ON THE PRESENCE OF SYMPTOMS AND ON WHETHER HYPONATREMIA IS ACUTE BUT LESS SO ON THE ACTUAL PLASMA [Na]

We will not deal here with the diagnosis and treatment of the underlying disease state.

TABLE 7–4. Drugs That May Cause High ADH Activity

Central Stimulation of ADH Release
Nicotine
Morphine
Clofibrate
Tricyclic antidepressants
Antineoplastic agents such as vincristine, cyclophosphamide
"ADH-Like" Agents
Oxytocin
Drugs Promoting ADH Action on the Kidney by Increasing Cyclic AMP
Oral hypoglycemics (e.g., chlorpropamide)
Methylxanthines (e.g., caffeine, aminophylline)
Analgesics that inhibit prostaglandin synthesis (e.g., aspirin, indomethacin)

TABLE 7–5. Treatment of Hyponatremia

Restrict Water Intake
Reduce ADH Levels
Correct the low "effective" circulating volume
NaCl if total ECF volume is low
Albumin (if its concentration is very low)
Improve myocardial function
Inotropic agents
Afterload reduction
Chronotropic agents (if needed)
Replace hormone deficiencies
Glucocorticoids
Thyroid hormone
Remove drugs that
Promote ADH release
Potentiate ADH action
Promote Water Loss Independent of ADH
Diuretics, giving electrolyte but not water replacement
Antagonists to ADH
Receptor blockers (new)
Lithium
Demeclocycline

Remove the Cause of ADH Release

With hyponatremia, the major potential physiologic stimulus for ADH release is a reduced "effective" circulatory volume (Table 7–2, page 175). Postural hypotension, low jugular venous pressure, and the absence of edema or ascites suggests Na deficiency. If the urine contains almost no Na and/or Cl, the patient has a nonrenal cause for Na loss or has taken diuretics in the past. The treatment is to restore the ECF volume with NaCl.

Also in the reduced "effective" circulating volume category are those patients with a reduced arterial volume without ECF volume contraction. Since total ECF volume is expanded, Na should not be given (it can aggravate congestive heart failure). Patients with poor myocardial function should receive specific therapy to improve cardiac performance. These measures alone may promote a Na and water diuresis. Patients with a very low albumin concentration may respond to exogenous albumin to initiate a water diuresis (not a long-term therapy).

In Na-depleted patients, give enough NaCl to reexpand the ECF volume promptly. The parameters to monitor are the disappearance of postural hypotension and a jugular venous pressure rise to several cm above the sternal angle. You can also gain information by measuring the urine [Na] and [Cl]; when Na and Cl are present in the urine, you have given enough saline (see page 185 for details of therapy). Once the ECF volume is normal (the urine contains NaCl), the stimulus for ADH release should disappear and a water diuresis should ensue (the half-life of ADH is only minutes). There is no control over the rapidity of the water diuresis unless ADH is given; sufficient Na should be given to continue to prevent the reoccurrence of ECF volume contraction.

The next reversible cause for augmented ADH action on the kidneys is the use of drugs that increase ADH release or its effect. A list is presented in Table 7–4, page 183, and these drugs are classified as to mechanism of action. If present, they should be discontinued (temporarily if necessary) and the patient observed for a water diuresis as before (be sure the urine contains NaCl, since more than one potential cause for ADH release may be present).

Aims in Therapy with Water Overload

- Raise the plasma [Na] 1 mM/h to shrink the ICF volume.
- Bring the plasma [Na] to close to 125 mM.
- Treat vigorously if the patient is symptomatic.
- Treat vigorously if the hyponatremia is acute.
- The severity of hyponatremia itself is not a critical determinant of aggressive therapy if the patient is asymptomatic.

Dangers in Therapy

- Rapid and/or severe overexpansion of the ECF volume.
- Too rapid correction or overcorrection of the water deficit may lead to central pontine myelinolysis.
- If ADH release is due solely to ECF volume contraction, rapid reexpansion of the ECF volume suppresses ADH release and causes a very large water diuresis and possibly central pontine myelinolysis. This can be avoided, if anticipated, by giving ADH once there is a water diuresis of 1–2 L.

Therapeutic Options

- Stop water intake (a *must*).
- Promote water loss (without Na loss). (Although this may be possible, it is not practical.)
- Promote water and Na loss but replace ongoing electrolyte losses. Usually a loop diuretic is used to increase these losses; the urine usually contains 40–100 mM Na and 5–30 mM K. These electrolyte losses should be replaced with *hypertonic* solutions to yield net water loss.
- Shift water from ICF to ECF with hypertonic saline if the patient needs or can withstand ECF volume expansion.

Calculation of Dose of Hypertonic Saline (an approximation)

- Since water moves across cell membranes to achieve isotonicity, the *TOTAL BODY WATER* is the denominator.
- Total body water is usually 60% of weight, but in severe hyponatremia, the ICF volume is considerably larger (the % rise is almost the same as the % decline in the plasma [Na]).
- The aim is to raise the [Na] by close to 1 mmol/L/h. The retention of 42 mmol of NaCl expands the ECF volume by close to 300 ml.
- Urine Na losses may be considerable.

Overall

- To raise the plasma [Na] from 110 to 125 mM requires 40 mmol NaCl/1 mM [Na] × 15 mM, or close to 600 mmol of positive balance for NaCl. This expands the ECF volume by 3–4 L if there is no excretion of the Na and water infused.

The final therapeutic option is to inhibit ADH release from the pituitary. Although in theory this can be achieved with ethanol, it does not work in practice. Phenytoin (dilantin) can be used to inhibit ADH release and may also be used diagnostically to confirm that the ADH is indeed released from the posterior pituitary (as compared with a peripheral site, e.g., a lung tumor), but it is not a first-line therapy or 100% effective (see Table 7–5, page 183).

Certain disorders in the SISADH category have a relatively short duration; when treated, the abnormal ADH release ceases. The most common of these are lung diseases such as pneumonia and tuberculosis, CNS disorders such as meningitis and encephalitis, and certain treatable space occupying lesions. These patients require water restriction until ADH release is normally controlled.

Restrict Water Intake

As long as water intake does not exceed water loss, the degree of hyponatremia does not get worse. Expect that the net nonrenal free water losses (sweat, respiration) will be close to 1 L per day. To these losses, we must add the urine water. If patients require IV medications, give them with saline, which provides less free water.

Promote Renal Free Water Loss

There are two major mechanisms to consider here:

1. Measures that lead to a cessation of ADH production or antagonize its action can produce a water diuresis.

2. A net free-water loss can be produced by making the urine contain a lower [Na] than ingested or infused saline. This process can be made more effective by increasing the volume of urine with a potent loop diuretic (furosemide). Each L of urine usually contains 40–100 mM NaCl and 5–30 mM KCl. These electrolyte losses should be replaced with as little water as possible (hypertonic saline). The net result is the loss of pure water (and some urea, which is not important). A reasonable goal is to raise the plasma [Na] by 1 mM/hr in symptomatic patients.

Drug Therapy

In the near future, the use of ADH antagonists may have therapeutic potential. These drugs allow a more liberalized water intake in patients with SISADH. Since ADH acts via

Central Pontine Myelinolysis. Central pontine myelinolysis, a symmetrical demyelination primarily in the base of the pons, was first described in 1959. Clinically, there is a wide spectrum of presentations going from the absence of symptoms to a devastating disorder characterized by confusion, agitation, and, eventually, flaccid or spastic quadriparesis. This may be accompanied by bulbar involvement. The diagnosis can be confirmed by CT scanning, including coronal views. There is no known treatment, and the mortality and morbidity rates are high.

With respect to etiology, central pontine myelinolysis occurs primarily in patients with disorders other than hyponatremia. However, it has also been observed after therapy for hyponatremia, and the incidence rises sharply if the hyponatremia is overcorrected. Hyponatremic patients in poor nutritional condition seem to have a much higher risk.

It seems prudent to raise the plasma [Na] at a rate of 1 mM/hr in symptomatic patients until 120 mM and/or or until the symptoms attributable to hyponatremia have abated; thereafter, full correction should be achieved over several days.

Symptomatic hyponatremia depends on the plasma [Na] and, more importantly, on the rate at which hyponatremia occurred. Thus, symptoms may be present at 120–125 mM [Na] during very acute hyponatremia, whereas symptoms may be absent at a [Na] below 110 mM with slowly progressive hyponatremia. Although there is no argument that severe symptomatic hyponatremia, especially if acute, needs to be treated promptly, asymptomatic hyponatremia needs to be treated much less vigorously.

the cyclic AMP system in the kidney, drugs that inhibit cyclic AMP formation such as lithium and demeclocycline can antagonize the effects of ADH. Only in rare circumstances (e.g., carcinoma of the lung causing SISADH) should demeclocycline be used. Lithium also has too many side effects to be considered as a front-line therapeutic agent.

Rapidity of Correction of Hyponatremia

- DO NOT OVERCORRECT HYPONATREMIA BECAUSE CNS DAMAGE MIGHT OCCUR
- RAISE THE PLASMA [Na] 1 mM/HR IF THE PATIENT IS SYMPTOMATIC

Since therapy for severe hyponatremia MAY be associated with a midbrain disorder due to cell swelling (central pontine myelinolysis; see page 187), do not overcorrect the hyponatremia. This CNS lesion should be suspected when the patient does not recover rapidly upon correction of hyponatremia or if there is a deterioration in the mental state following therapy. There is a controversy concerning the rapidity of correction of hyponatremia; several review articles on this subject are provided in the reference list at the end of the Na and water chapters.

Suggested Readings

Anderson B: Regulation of water intake. Physiol Rev 58:582, 1978.
Anderson RJ: Hospital-associated hyponatremia. Kidney Int 29:1237–1247, 1986.
Arieff AI: Hyponatremia, convulsions, respiratory arrest, and permanent brain damage after elective surgery in healthy women. N Engl J Med 314:1529–1535, 1986.
Ayus JC, Krothapalli RK, and Arieff AI: Treatment of symptomatic hyponatremia and its relation to brain damage. New Eng J Med 317:1190–1195, 1987.
DeFronzo RA, Goldberg M, and Agus ZA: Normal diluting capacity in hyponatremic patients. Reset osmostat or a variant of the syndrome of inappropriate antidiuretic hormone secretion. Ann Intern Med 84:538, 1976.
duVigneud V: Hormones of the posterior pituitary gland: oxytocin and vasopressin. In The Harvey Lectures, 1954–55, New York, Academic Press, 1956.
Goldfarb S and Agus ZA: Mechanism of the polyuria of hypercalcemia. Am J Nephrol 4:69–76, 1984.
Jamison RL: A patient with polyuria and hyponatremia. Kidney Int 24:256–267, 1983.
Miller M and Moses AM: Drug-induced states of impaired water excretion. Kidney Int 10:96, 1976.
Narins RG: Therapy of Hyponatremia: Does haste make waste? N Engl J Med 314:1573–1575, 1986.
Raymond KH, Davidson KK, and McKinney TD: In vivo and in vitro studies of urinary concentrating ability in potassium-depleted rabbits. J Clin Invest 76:561–566, 1985.
Robinson AG: Disorders of antidiuretic hormone secretion. Clin Endocrinol Metab 14:55–88, 1985.
Rose BD: Hyponatremia. Med Gr Rds 4:75–83, 1986.
Schrier RW, Berl T, and Anderson RJ: Osmotic and nonosmotic control of vasopressin release. Am J Physiol 236:F321, 1979.
Schwartz WB, Bennett W, Curelop S, et al: Syndrome of renal sodium loss and hyponatremia probably resulting from inappropriate secretion of antidiuretic hormone. Am J Med 23:529–542, 1957.
Sterns RH: Severe symptomatic hyponatremia: treatment and outcome. Ann Intern Med 107:656–664, 1987.
Sterns RH, Riggs JE, and Schochet S Jr: Osmotic demyelination syndrome following correction of hyponatremia. N Engl J Med 314:1535–1542, 1986.
Zerbe RL, Stropes L, and Robertson GL: Vasopressin function in the syndrome of inappropriate antidiuresis. Annu Rev Med 31:315, 1980.

8

HYPERNATREMIA

THE BIG PICTURE

In all patients with hypernatremia, the ICF volume is contracted. The brain is most susceptible, and a CNS hemorrhage might ensue.

Na gain is rarely responsible for hypernatremia; detect it by finding an expanded ECF volume.

Hypernatremia is almost always due to water loss.

There are 2 major expected responses to water loss: thirst and the excretion of the *minimum* volume of *maximally* concentrated urine. Failure to detect these responses should signal the site of the lesion.

The urine osmolality differentiates the 3 major causes: Diabetes insipidus (large volume of hypoosmolar urine), osmotic or pharmacologic diuresis (large volume of slightly hyperosmolar urine) and nonrenal water loss without water intake (minimum volume of maximally hyperosmolar urine).

Hypernatremia is not a specific disease: look for the cause and treat the underlying disease.

NORMAL VALUES RELEVANT TO HYPERNATREMIC PATIENTS

[Na]

The actual plasma [Na] is 152 mmol/kg H_2O. However, if measured per L of plasma, the plasma [Na] is 140 mM (because this L contains 6–7% nonaqueous volume [lipid, protein]). The normal range of plasma [Na] is 136–144 mM. If blood lipids or protein is excessively high, the laboratory values for the plasma [Na] are much lower than the actual Na:H_2O ratio. If the plasma [Na] is measured using an Na-selective electrode or a conductance method (i.e., measuring the Na:water ratio) on an undiluted sample, the normal value is 152 mM; notwithstanding, the lab will "back-calculate" this value and report it as 140 mM. There is no "correction factor" needed for hyperlipidemia or hyperproteinemia in these cases.

Urine Osmolality (or Specific Gravity)

There is *no normal value* for urine volume and osmolality! The kidney responds to a change in tonicity of body fluids, i.e., excretes the difference between needs and intake. If no water was taken in, the kidney should excrete the minimum volume it can (400–800 ml/day), with the maximum osmolality (1200 mOsm/kg H_2O, specific gravity [SG] 1.030); a lower osmolality is observed in the presence of renal disease.

Thirst

Thirst is initiated by a rise in the plasma [Na] of 2 mM. Look for this response in all hypernatremic patients.

PHYSIOLOGY CAPSULE FOR Na AND H_2O

(see Chapter 7, page 166)

Treatment of hypernatremia has two components: First, stop free-water loss if possible and, second, administer free water. Hypotonic saline and glucose in water are the IV solutions. Do not give glucose faster than the patient can metabolize it.

ILLUSTRATIVE CASE

A patient developed acute meningitis; she showed confusion at this time. The plasma [Na] was 140 mM, the ECF volume was slightly low, and the urine volume was low. The patient had a series of convulsions and was treated with a large dose of the anticonvulsant phenytoin (Dilantin). Over the next 8 hr, she had the abrupt onset of polyuria without complaining of thirst. Physical exam revealed slight ECF volume contraction and a weight loss of greater than 10 lb. The plasma [Na] rose to 157 mM and urine osmolality was 100 mOsm/kg H_2O.

Questions (Answers on page 193)

1. Of what significance is the fact that the ECF volume is slightly contracted?
2. Why is thirst not a feature in this patient?
3. What significance is the loss of body weight?
4. What is the significance of the polyuria? Can any inference be made from the urine osmolality and the acuteness of the onset?
5. What should your therapy be?

ADDITIONAL ILLUSTRATIVE CASES

You are confronted with 3 patients; each has hypernatremia (154 mM). One patient drank sea water, another lost pure water, whereas the third lost hypotonic saline (sweating or via a diuretic).

Questions (Answers on page 193)

6. What parameter would distinguish these three patients?
7. What is the major threat(s) to life in each case?

Answers

1. Hypernatremia is due to Na gain or water loss from the ECF. The reduced ECF volume indicates that the hypernatremia in this case is due to water loss, not Na gain.

2. Hypernatremia elicits a thirst response unless a CNS lesion is present (generalized CNS problem [confusion], a specific lesion involving the thirst center, or a communication problem). This patient was confused.

3. The loss in weight indicates that the cause of hypernatremia is water loss from the body (not a shift into muscles).

4. Polyuria suggests a problem of either diabetes insipidus or the presence of a diuretic. Since the urine is quite hypoosmolar in the face of hypernatremia, the patient has diabetes insipidus. Furthermore, such a low osmolality and the acuteness of onset both suggest that central rather than nephrogenic diabetes insipidus is the diagnosis. In this case, the meningitis and/or the phenytoin could have caused a low ADH release. This impression can be confirmed by an appropriate response to exogenous ADH.

5. Therapy has 2 major aims: First, stop water loss by giving ADH; second, water must be given either as half-normal saline [danger is ECF volume expansion] or glucose in water, provided that hyperglycemia is not present. Do not give more than 0.3 L/hr (dangers are hyperglycemia and a glucose-induced osmotic diuresis causing additional water loss).

6. *Distinguishing Features:* The high Na:H_2O ratio of 154 mM indicates that the ICF volume is contracted to the same degree in all cases. However, the patient who ingested sea water has ECF volume expansion. In contrast, in the patient who lost hypotonic saline, the Na loss produces a reduced ECF volume. The loss of water per se causes only a small ECF volume change.

7. *Major threats to life:* In each case, the major organ at risk is the brain, i.e., for hemorrhage. In addition, in the patient who drank sea water, there is the potential danger of congestive heart failure. In the patient who lost hypotonic saline, shock could ensue.

CLINICAL APPROACH TO THE PATIENT WITH HYPERNATREMIA

- HYPERNATREMIA IS ALMOST ALWAYS DUE TO WATER LOSS; MANY PATIENTS ALSO HAVE SOME Na LOSS
- QUESTIONS TO BE ANSWERED WITH HYPERNATREMIA:
  1. WHAT IS THE ECF VOLUME?
  2. HAS THE BODY WEIGHT CHANGED?
  3. IS THE THIRST RESPONSE NORMAL?
  4. IS THE RENAL RESPONSE NORMAL?

What is the ECF Volume? Hypernatremia due to Na gain rarely occurs. It is characterized by ECF volume expansion, which is easily detected on physical examination. All other cases of hypernatremia are due to water loss (i.e., no ECF volume expansion).

Has the Body Weight Changed? Very rarely, water is lost from the ECF owing to a shift into the ICF (e.g., rhabdomyolysis). Only in this latter case is hypernatremia accompanied by a decrease in the ECF volume *without* a loss of body weight.

Is the Thirst Response to Hypernatremia Normal? A 2% rise in plasma tonicity provokes a powerful urge to drink; therefore, hypernatremia should be accompanied by thirst. Failure to drink could occur if the patient could not get water (confusion, paralysis, etc.). The absence of a thirst response should alert the clinician to suspect a generalized or localized CNS lesion.

Is the Renal Response to Hypernatremia Normal? The appropriate renal response to pure water loss is the excretion of urine with the *highest* osmolality that the kidney can achieve (not just above isotonicity) and a urine volume that is the *minimum* value that can be excreted (about 0.5 L/day). Deviations from this response signal an ADH or renal problem (a renal disorder or osmotic load).

RENAL RESPONSE TO HYPERNATREMIA
- URINE VOLUME 20 ml/hr (0.5 L/day)
- URINE OSMOLALITY 1000 + mOsm/kg H_2O

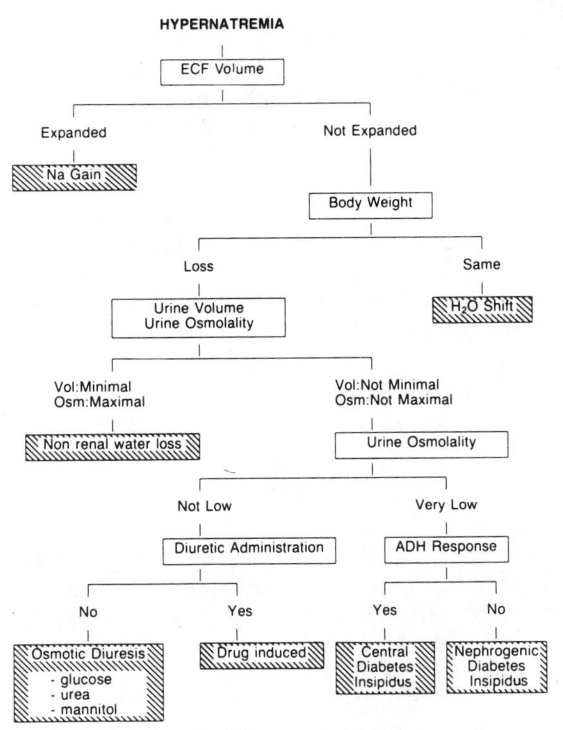

Flow Chart 8–A. Approach to hypernatremia. The open boxes represent the variables that, when evaluated, permit one to progress towards the diagnosis. The diagnostic groups are shown in the shaded boxes.

ETIOLOGY OF HYPERNATREMIA

> HYPERNATREMIA IS *VERY* RARELY DUE TO Na GAIN.

Na Gain. Na gain is observed in 4 major clinical situations: first, the patient given a hypertonic Na salt intravenously ($NaHCO_3$) in a cardiac arrest or hypertonic NaCl administration); second, the ingestion of sea water; third, the replacement of sugar with salt in the pediatric formula; fourth (and most common), the replacement of hypotonic Na loss with isotonic saline (i.e., during the treatment of diabetic ketoacidosis because the urine in a patient with an uncomplicated osmotic diuresis should have a [Na] of close to 50 mM). If hypernatremia is severe, confusion or convulsions may be present. In all but the fourth case, the ECF volume is expanded.

Treatment in the first 3 cases is to increase Na loss with a diuretic and to give water by mouth if possible. Recall that there is a limited amount of glucose in water that can be given safely intravenously.

Question (Answer on page 197)

7. What is the cause of hypernatremia in the following case? The plasma [Na] rose to 180 mM overnight in a confused patient. He received no medications. His body weight had not changed; the ECF volume was contracted. His blood sugar was 90 mg/dl (5 mM), the urine volume was extremely low, and the urine osmolality was close to 1200 mOsm/kg H_2O.

Water Loss as the Cause of Hypernatremia

Nonrenal Water Loss. Water loss via the respiratory tract and skin (perspiration, a hypotonic solution with a lower [Na] when volumes are high) is, on the average, just under 1 L/day. Losses in perspiration can increase dramatically in hot environments and with exercise. Similarly, water losses in patients who are febrile and hyperventilating are higher. Should the patient have a generalized or localized CNS lesion involving the thirst mechanism or the inability to obtain water, hypernatremia may develop because water

Answer

7. The patient did not receive Na salts, and the ECF volume was contracted; therefore, hypernatremia was not due to Na gain. Although gastrointestinal secretions have a [Na] that is less than 140 mM, there was no clinical evidence to support the accumulation of very large volumes of these fluids. Since there was no change in body weight, hypernatremia was not the result of water loss from the body (water loss should have caused a weight loss of more than 10 lb), and the renal response to hypernatremia was normal (low volume of maximally concentrated urine). Hence, the rise in [Na] was due to a shift of water from the ECF to another compartment.

Since the serum [Na] increased by 40 mM (140 to 180 mM), the osmolality rose by almost 30%; this could occur if almost 8 L of water shifted from the ECF (plus an additional water shift from the ICF of other organs that were not involved in the pathologic condition) to the ICF of the diseased organ. Judging by the relative masses of individual organs, water must have accumulated in muscle.

For water to shift into cells, the number of particles in these cells must have increased (and Na did not accumulate in these cells). Rhabdomyolysis, with a breakdown of intracellular muscle macromolecules into a larger number of smaller particles that were retained in the ICF of muscle, could result in an increased tonicity of these cells.

From a clinical viewpoint, it is essential to realize that although water shifted *into* muscle cells, it shifted *out* of other normal cells (e.g., brain cells). Shrinkage of these cells could be life-threatening. Therefore, therapy should orient around defense of both the ECF volume and the ICF volume of nonmuscle tissues (e.g., brain cells), and half-normal saline is the best solution for these purposes. A further expansion of the ICF of muscle should be anticipated with therapy, and specific steps may be necessary to cope with this (i.e., decompression).

loss is not matched by water intake. Another clinical group in which hypernatremia due to nonrenal water loss is not uncommon is infants who cannot specifically complain of thirst. In the adult cases, the urine is of *minimum* volume (400–800 ml/day) and of *maximum* osmolality (~1200 mOsm/kg H_2O).

Renal Water Loss. Two distinct types of renal problems (separable by measuring the urine osmolality) can lead to Na-free water loss and thereby to hypernatremia. It is well known that patients with hypernatremia who excrete large volumes of hypoosmolar urine suffer from diabetes insipidus. If the urine volume can be lowered and the urine osmolality raised appreciably in such a patient following the administration of a potent ADH preparation, the defect is in the synthesis of or release of ADH from the brain; this is termed *central diabetes insipidus* (Table 8–1, page 199). Alternatively, should the urine remain hypoosmolar or close to isosmolar and the volume is large after ADH administration, *nephrogenic diabetes insipidus* is present. These 2 conditions are discussed below.

CENTRAL DIABETES INSIPIDUS (Table 8–1, page 199)

> Central diabetes insipidus occurs in patients with a CNS lesion. The plasma [Na] is high and polyuria is present. With ADH administration, the urine volume declines and osmolality rises.

A rise in the plasma [Na] (but not hyperosmolality due to hyperglycemia or a high urea concentration) stimulates the hypothalamic osmoreceptor and leads to augmented ADH synthesis in the paraventricular and supraoptic nuclei. This ADH is then transported by axonal flow to the posterior pituitary. Hence, a lesion at any of these sites produces central diabetes insipidus. The most common causes for a lesion here are trauma (especially basal skull fractures), infections, space-occupying lesions, and neurosurgical procedures.

Selective removal of the posterior pituitary usually produces only a transient central diabetes insipidus (it seems that ADH can be released from the hypothalamic neurons that synthesize it). In almost every case, the onset of symptoms of polyuria and polydipsia are *abrupt*. In central diabetes insipidus due to surgery or trauma, *3 phases of response* may be seen:

TABLE 8–1. Etiology of Central Diabetes Insipidus

Trauma (especially basal skull fractures)
Neurosurgery (hypophysectomy, other)
Space-Occupying Lesions
Neoplasm
PRIMARY
Craniopharyngioma, pineal cyst, pituitary tumors
SECONDARY
Metastatic
Granuloma
Sarcoid
Histiocytosis X
Infection (meningitis, encephalitis)
Vascular (aneurysm)
Posthypoxia
Drugs Interfering With ADH Release (e.g., phenytoin)
Idiopathic (may be familial)

Stage I: Initial polyuria for up to a couple of days (due to inhibition of ADH release)

Stage II: Antidiuresis for up to a few days (release of stored ADH from the degenerating gland)

Stage III: Permanent central diabetes insipidus

The important aspect of recognizing this triad is to avoid the misconception that the disorder was only temporary, and thus one should continue to observe this patient closely.

The diagnosis of central diabetes insipidus is usually easy to confirm. The patient has a CNS problem together with a history of polyuria and polydipsia (preferring cold liquids). Physical examination may help identify the underlying disorder. Laboratory exam shows a high-normal or high plasma [Na]. The urine volume is high (3–20 L/day, unless the GFR is low) and the osmolality is inappropriately low. The diagnosis is confirmed when ADH administration results in a prompt rise in the urine osmolality (to above that of plasma but not necessarily to maximum values because a period of time is required for the osmolality in the medullary interstitium to become elevated).

Anything that diminishes the delivery of filtrate to the distal nephron can lower free-water excretion. Hence, polyuria is less dramatic in the presence of ECF volume contraction or anterior pituitary resection (loss of the hemodynamic benefits of cortisol and thyroid hormone).

In close to 50% of cases of central diabetes insipidus, no specific cause is identified (this is called *idiopathic central diabetes insipidus*). The majority of the other cases are caused by head trauma, neurosurgery, and neoplasms.

Questions (Answers on page 201)

8. A patient had a previous stroke with aphasia. All lab tests were normal. She was transferred to a chronic care facility, where tube feeding was instituted; no drugs were given. She was readmitted to the hospital several weeks later, at which time her ECF volume was contracted. The plasma [Na] was 160 mM, the urine osmolality was 450 mOsm/kg H_2O, and the urine volume was 3–4 L per day. The urine glucose was negative. What is your diagnosis, and how would you confirm it?

9. How may hypokalemia and hypercalcemia influence the urine volume?

Answers

8. Since hypernatremia was associated with ECF volume contraction, it was due to water loss. The water loss was renal, as the patient was polyuric; this was not a water diuresis, since the urine osmolality was high. Therefore, an osmotic diuresis is the most likely cause (no drugs given). In the absence of glucosuria, a urea-induced diuresis should be suspected (to confirm, measure the urine [urea]; it should be at least 250 mM). The urea load was probably due to the high protein feeding.

Note the absence of (or failure to communicate) thirst, owing to the previous stroke with resulting aphasia.

In addition, an osmotic diuresis leads to Na loss—hence, the prominent ECF volume contraction.

9. The mechanisms whereby hypokalemia produces polyuria are multiple. Circulating levels of ADH are normal. Although ADH-induced cyclic AMP generation is reduced, the physiologic response to ADH appears to be normal. Medullary hyperosmolality is decreased with hypokalemia, and impaired NaCl reabsorption in the thick ascending limb has recently been shown. These effects require a large (200+ mmol) K deficit; reversal with K therapy may take many weeks.

With hypercalcemia, levels of at least 11 mg/dl (2.75 mM) are required to induce this defect. Again, mechanisms producing nephrogenic diabetes insipidus are multiple. Circulating levels of ADH are normal during dehydration; however, there is evidence that the renal response to ADH is impaired. Hypercalcemia stimulates prostaglandin E_2 production, and this compound inhibits both the hydroosmotic effects of ADH and the stimulating effect of ADH on NaCl reabsorption in the medullary thick ascending limb of the loop of Henle. Hypercalcemia also causes a reversible decline in the GFR. Hence, the major effect seems to involve a lower medullary hyperosmolality.

Primary polydipsia contributes to polyuria. It has been shown with both hypokalemia and hypercalcemia, but the derangement in thirst does not cause nephrogenic diabetes insipidus.

NEPHROGENIC DIABETES INSIPIDUS (Table 8–2, page 203). This disorder can be divided into two major categories: first, those in which ADH fails to increase collecting duct water permeability, and thus there is not osmotic equilibrium between the hypertonic medulla and the hypotonic luminal fluid. Second, there is a group of diseases in which there is a loss of medullary hypertonicity due to a major medullary interstitial defect or infiltrate.

The typical urine output in these patients is only 4–5 L/day, and the urine is generally close to isotonic. This differs from the extreme polyuria and maximally dilute urine of central diabetes insipidus. By definition, these patients do not have changes in urine volume or osmolality when ADH is given (see page 207 for more details).

SYMPTOMS DUE TO HYPERNATREMIA

The only symptom directly related to hypernatremia is mild confusion. However, severe hypernatremia can lead to CNS dysfunction and, ultimately, to coma and hemorrhages (subarachnoid or intracerebral). The exact [Na] that can produce symptoms is lower if the onset of hypernatremia is acute (160 mM seems to be close to the range in which symptoms are common). In contrast, symptoms may be absent if the hypernatremia develops gradually.

The polydipsia of central diabetes insipidus is associated with a strong preference for ice-cold liquids, a preference that is not common in other polyuric states. Some patients can suppress frequency of voiding and thus urinate very large volumes (close to 1 L) less often; they may develop a dilated bladder, hydroureter, and even hydronephrosis.

TREATMENT OF WATER DEFICIT

1. STOP ONGOING WATER LOSS.
2. REPLACE DEFICIT SLOWLY, BY ORAL ROUTE IF POSSIBLE.

IF YOU MUST GIVE INTRAVENOUS FREE WATER, USE HYPOTONIC SALINE PLUS A LIMITED QUANTITY OF GLUCOSE IN WATER.

TABLE 8–2. Etiology of Nephrogenic Diabetes Insipidus

Compromised ADH-Induced Cyclic AMP Rise
- Lithium
- Demethylchlortetracycline
- Congenital nephrogenic diabetes insipidus

Loss of Medullary Hypertonicity
- ***Renal Medullary Pathology***
  - Infiltrations (amyloid, etc.)
  - Infections (pyelonephritis)
  - Drug-induced
  - Hypoxic damage (sickle-cell anemia)
  - Obstructive uropathy
- ***Compromised Medullary Hypertonicity***
  - Loop diuretics
- ***Generalized Kidney Disease***
  - Polycystic disease
- ***Electrolyte Abnormalities***
  - Hypokalemia
  - Hypercalcemia

Idiopathic

The major objectives of therapy for patients with hypernatremia due to water loss are to treat the underlying medical problems, to minimize ongoing water losses, and to replace the water deficit.

Stop the Water Loss. The clinician must curtail large ongoing water losses if possible. The problem of ADH deficiency can easily be rectified with appropriate replacement therapy. If an osmotic diuresis is present, the addition of the osmotic agent must be stopped, if possible, and ongoing Na and K losses be replaced.

Replace Water Deficit. Too-rapid replacement of the water deficit should be avoided (> 1 mM fall in $[Na^+]$/hr). The water deficit should be calculated (see page 205), and only a portion (perhaps one half of this deficit plus ongoing water losses) should be replaced over the next 12 hr. Total correction should occur over several days. The oral route of water replacement is preferable if the patient is conscious and alert. This avoids the problems of hyperglycemia and glucosuria associated with intravenous glucose in water administration. The advantages and disadvantages of the intravenous solutions used for free water administration are discussed in the following paragraphs (glucose in water and 75 mM saline [half-normal]).

Glucose in Water (D_5W). D_5W appears to have ideal properties to deliver free water intravenously; i.e., it is isotonic (thus avoiding hemolysis), and the osmoles (glucose) disappear as a result of metabolism. However, it is not generally appreciated that glucose cannot be metabolized at very rapid rates (see page 205).

Half-Normal Saline. As mentioned earlier, it is important to stop water losses early because it is difficult to administer large volumes of free water safely. Water deficits should be replaced with "half-normal" saline and small quantities of D_5W (a maximum of 0.3 L/hr). Care should be taken to avoid a serious Na overload. Since each L of half-normal saline has 500 ml of free water that distributes throughout total body water, only 333 ml enters the ICF. In quantitative terms, if the ICF volume is 30 L, then 11 ml of free water will be added to each L of ICF (the brain contains less than 1 L of ICF). More free water can be administered if even more hypotonic saline solutions are infused (0.25% or 0.33%).

DIFFERENTIAL DIAGNOSIS OF POLYURIA

Three major groups can be identified on the basis of urine osmolality: hyperosmolar, isosmolar, and hypoosmolar.

Hyperosmolar. For a large volume of hyperosmolar urine to be excreted, there must be a source of these osmoles. If the urine osmolality is, say, 400 mOsm/kg H_2O, and 5 L of urine are produced/day, 2000 mOsm must be identified (normal = less than half this quantity). If the urine contains much less than 150 mM of Na + K, then look for an osmotic diuretic (close to 300 mM of glucose or urea). A high urea excretion occurs with high protein feeding, trauma, gastrointestinal hemorrhage, a catabolic state (neoplasm or sepsis), or a high blood urea (recovery from obstructive uropathy or renal failure—other factors also operate to cause polyuria in these conditions). It is not necessary to administer ADH to help establish the diagnosis in this group of patients.

It should be emphasized that each L of urine excreted during an osmotic diuresis is expected to contain close to 50 mM Na and 50 mM K. If the administration of Na and K was less than 150 mmol/day, deficiency states would ensue, leading to ECF volume contraction and hypokalemia.

Isotonic Polyuria. The major differential diagnosis is summarized in Table 8–2, page 203 (nephrogenic diabetes insipidus). These patients do not show a rise in urine osmolality of greater than 50% following ADH administration. It is possible that rare patients with central diabetes insipidus can have isosmolar urine (those with marked ECF volume contraction that lowers urine flow rate appreciably and those with an incomplete defect in ADH production—low but not absent ADH levels). In both of these cases, you should expect a rise in urine osmolality of at least 50% with ADH administration.

Hypoosmolar Urine. Almost all hypernatremic patients with very hypoosmolar urine have central diabetes insipidus; they excrete hyperosmolar urine when ADH is given. An exception to this rule is certain patients with nephrogenic diabetes insipidus (lithium or demethyltetracycline therapy, congenital nonresponsiveness to ADH); they do not have an appreciable rise in urine osmolality after ADH administration.

Distilled Water. If a patient is severely hypertonic, is in congestive heart failure, is severely hyperglycemic, and cannot tolerate dialysis or oral water therapies, it may be necessary to administer water intravenously. This water should be given by central vein, and several L may be administered/day by this route.

Selected Reading

Bjorntorp P and Sjostrom L: Carbohydrate storage in man: speculations and some quantitative considerations. Metabolism 27:1853–1865, 1978.

Marsden PA and Halperin ML: Pathophysiologic approach to patients presenting with hypernatremia. Am J Nephrol 5:229–235, 1985.

Robertson GL: Abnormalities of thirst regulation. Kidney Int 25:460–469, 1984.

THE QUANTITATIVE ASPECTS OF GLUCOSE METABOLISM

One L of D_5W contains 278 mmol (50 g) of glucose. The total glucose content of the body (5 mmol/L × 14 L ECF + 4 L ICF) is only 90 mmol (16.2 g) when the blood glucose concentration is 5 mM (90 mg/dl). Hence, without metabolism, 1 L of D_5W increases the blood glucose concentration by 3-fold, and, because several L of D_5W are required for water replacement during severe hypernatremia, this could greatly aggravate a life-threatening hyperosmolar state.

Approximately 0.5 g of glucose can be metabolized/kg of body weight/hr, providing that a fat-derived fuel is not available for oxidation. However, during fasting, when ketone bodies are oxidized, less than 0.1 g of glucose/kg of body weight is metabolized to CO_2. Therefore, oxidative metabolism of glucose in major organs is close to 7 g/hr. Since the liver does not oxidize large quantities of circulating glucose to CO_2, and since glucose conversion to glycogen in liver and muscle is rather limited in acutely ill patients, the maximum rate of glucose metabolism could be 10 g/hr. Taken together, these factors should alert the clinician to the life-threatening danger of profound hyperglycemia consequent to large D_5W infusions. We therefore advise that the rate of administration of D_5W not exceed 0.3 L/hr.

CALCULATION OF WATER DEFICIT

The simplest approach here is to obtain quantitative estimates of ECF and ICF volume deficits.

ICF Volume. The ICF volume is inversely related to that of the plasma [Na]. For example, a 10% rise in [Na] (154 mM) signals a 10% decline in ICF volume (2.8 L, 10% of 28 L). Actually, one must infuse 50% more than this volume (4.2 rather than 2.8 L in this example) so that the tonicity and volume of the ICF and ECF are both returned to normal.

ECF Volume. The ECF volume is assessed clinically, and deficits over and above those due to the water deficit should be replaced with the equivalent of isotonic saline (of course, twice the volume of half-normal saline is better because some of the water deficit will be replaced at the same time).

9

POTASSIUM PHYSIOLOGY

THE BIG PICTURE

***K Distribution*. 98% of body K is in the intracellular fluid (ICF). In muscle, the ICF [K] is close to 150 mM and this is 35–40 times higher than the [K] in the extracellular fluid (ECF) (4 mM). This [K] gradient is the result of K transport into cells by the NaKATPase and of K diffusion out of the cell, passively, down the [K] gradient. These events generate the resting membrane potential. During hypokalemia or hyperkalemia, the change in ECF [K] is proportionally larger than that in the ICF. A rise in the ECF [K] can depolarize cells and lead to cardiac arrhythmias. During hypokalemia, cells tend to hyperpolarize, and this can also lead to cardiac arrhythmias.**

***Time Course of Defense of the ECF [K]*. Adults consume daily about as much K as they have in their ECF (1 mmol/Kg body weight). After a meal, virtually all of absorbed K is first shifted into cells (time period**

A QUANTITATIVE ANALYSIS OF K DISTRIBUTION

Absorb 70 mmol K

Rapid (minutes): With no change in the resting membrane potential, 69 of the 70 mmol of K enter the ICF.* The distribution of K into the ICF is even greater if hormones (β_2-adrenergics, insulin) act to increase the resting membrane potential. In response to an elevation in the ECF [K], albeit small, aldosterone is released promptly from the adrenal cortex.

Excrete 70 mmol of K in the Urine

Slow (hours): Aldosterone acts on kidneys, but a lag period is required. Aldosterone acts on the late distal convoluted tubule and cortical collecting duct; the net effect is a rise in the luminal [K] by the following mechanisms:

1. Lumen Na^+ channel is opened;
2. NaKATPase is activated;
3. Lumen K^+ channel is open;
4. Luminal membrane Cl conductance falls.

---

*Movement of 69 mmol of K into cells

Given:

1. ICF [K] exceeds that in the ECF by 37.5-fold.
2. ICF volume is 2-fold higher than that of the ECF. Thus, the ICF has 75–80-fold more K than the ECF.

Therefore: If no change in the resting membrane potential (RMP):

1. Raise the ECF K by 1 mmol and the ICF K by 69 mmol.
2. New ECF [K] = 61 mmol (4 mM × 15 L plus 1)/15 L = 4.1 mM.
3. New ICF [K] = 4569 mmol (150 mM × 30 L plus 69)/30 L = 152 mM.
4. No hormone action is needed to increase the resting membrane potential to get the above K distribution.

is in minutes); thereafter, renal K excretion occurs (time period is in hours).

An individual with normal K metabolism can consume 1 mmol K/kg of body weight in 1 meal and not be threatened by severe hyperkalemia, even if kidney function is suddenly lost.

Renal K Excretion. **K excretion is the product of the urine [K] and the urine flow rate. Aldosterone raises the [K] in the cortical distal nephron, whereas Na and water balance and hemodynamic considerations determine the urine flow rate.**

Quantitative analysis and time course for K handling are provided on page 209.

Questions (Answers on page 211)

1. What is the time course for, and relative importance of, the shift of K into the ICF and the renal excretion of K as a defense against postprandial hyperkalemia?
2. What are the most important factors regulating urinary K excretion?
3. What is the maximum rate of K excretion in 24 hr? (Given: plasma [K] 4 mM, transtubular [K] ratio 10, 10 L of filtrate reach the cortical collecting duct, and no medullary K reabsorption.)
4. How much will the plasma [K] change when 1 L of pure water is shifted from the ICF to the ECF? (Assume a constant membrane potential—28 L ICF, 14 L ECF.)
5. Will a muscle relaxant that acts by depolarizing cell membranes cause a change in the plasma [K]?

FORCES ACTING AT THE ECF-ICF INTERFACE (Table 9–1, page 217).

- K^+ PUMPED INTO CELLS: NaKATPase
- K^+ DIFFUSE OUT OF CELLS: RESTING MEMBRANE POTENTIAL CLOSE TO −90 mVOLTS IN MUSCLE

Answers

1. Since the kidney takes several hours to excrete the K load, an immediate shift of K into cells is the first critical line of defense against hyperkalemia. The kidney mechanisms must ultimately work for K homeostasis.

2. We can subdivide these into 2 major groups and 2 permissive factors:

a. Aldosterone, which increases the [K] in the lumen of the cortical collecting duct.

b. Factors that increase the volume of fluid delivered to the "cortical distal nephron."

c. The permissive factors are distal Na delivery and ADH for distal luminal membrane K conductance.

3. The maximum rate of K excretion can be deduced as follows:

The luminal fluid [K] is 10 × the plasma value, or 40 mM at equilibrium, if aldosterone is fully active for the 24-hr period.

The volume of fluid in the end of the cortical collecting duct is 10 L/day.

Therefore, the K excreted is 400 mmol/day (10 L/day × 40 mM).

4. Less than 0.2 mM. Should the pure water shift per se not change the resting membrane potential, then the shift of 1 L of water from the ICF to the ECF of a 70 kg person would result in a very small rise in the ECF [K] because the ICF volume would decrease from 28 to 27 L (i.e., by 4%). Thus, the ICF [K] would rise by this amount, since there is no appreciable change in K content. Therefore, at a constant resting membrane potential, the [K] in the ECF rises by an equivalent percentage, i.e., less than 0.2 mM (the actual quantity of K shifted is very small).

5. Anything that depolarizes cell membranes decreases the transmembrane [K] ratio. Therefore, there will be a shift of K from the ICF to the ECF when this membrane potential falls, which leads to hyperkalemia. This change is usually small unless a major depolarization occurs.

The ICF [K] depends mainly on the number of anions (largely organic phosphates) that are restricted to the ICF compartment. This [K] can be altered by replacement of ICF K^+ by another cation, maintaining electroneutrality. The cations that replace K^+ in the ICF are Na^+ and H^+; thus, this cation-cation shift not only changes the ratio of the [K] in the ICF:ECF, but it may also produce pH and/or volume changes in the ICF.

> NET K MOVEMENT ACROSS THE CELL MEMBRANE REQUIRES:
> - EXCHANGE OF K FOR Na^+ AND/OR H^+
> - SIMULTANEOUS MOVEMENT OF K^+ AND AN ICF ANION

Role of Intracellular Anions

The number of positive and negative charges in a compartment are equal. In the ICF, the negatively charged molecules do not tend to cross cell membranes (mainly organic phosphates, DNA, RNA, ATP, creatine phosphate). Intracellular proteins do not make a major net contribution to this anionic charge (see page 213).

The intracellular anionic composition is normally relatively constant (although the specific phosphate compounds differ from organ to organ). However, loss of organic phosphates, as can occur in specific disease states (e.g., diabetic ketoacidosis), leads to a parallel loss of "accompanying" ICF K^+. Hyperkalemia will result unless this K is excreted in the urine (it usually is because there is an osmotic diuresis; see page 280). These effects of ICF anion loss during diabetic ketoacidosis stand out in contrast with the effects of muscle wasting due to poliomyelitis, for example. A patient with loss of muscle mass has a low total body K *content* but a normal ICF [K] and a normal [K] gradient across the remaining cell membranes.

The need to provide ICF anions to restore ICF K^+ explains why phosphate plus K must be given in therapy for diabetic ketoacidosis or for severe anemias (Fig. 9–1, page 213). Similarly, when cell growth (anabolism) resumes, the patient needs not only K supplements to avoid hypokalemia but also

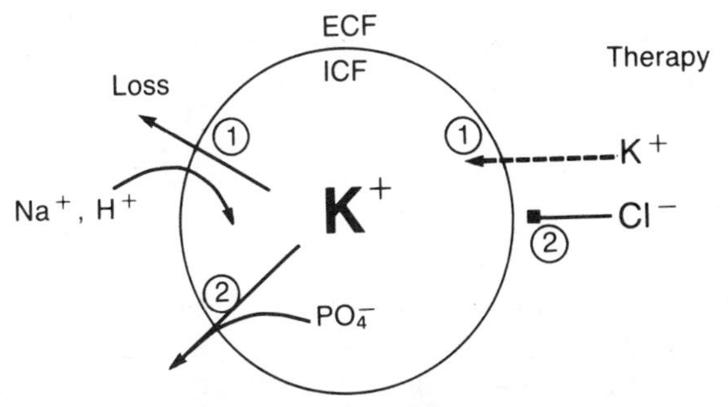

Figure 9–1. The effect of KCl administration on a patient with K deficiency. As shown on the left-hand side, K can be lost from the ICF by two mechanisms: (1) a cation-cation pathway and (2) a loss of K plus an ICF anion (phosphate).

During therapy (right-hand side), since the entire K deficit was not produced by the cation-cation exchange (mechanism 1), KCl *alone* will not correct the entire ICF K deficit because C1 is predominantly an ECF anion. To replace the K deficit in the ICF, some K must be given along with an anion, which distributes in the ICF (mechanism 2, phosphate, not Cl^-).

Net Charge on Intracellular Proteins. At the ICF pH, the anionic charges on proteins are on the C-terminal carboxyl group, glutamate and aspartate. The positive charges on ICF proteins are on the N-terminal amino group, lysine, arginine and less than ½ of the histidines. Simple arithmetic shows that the anionic groups do not exceed the cationic residues. Hence, proteins, in general, are not likely to be major ICF anions unless very marked changes in their pK occur in specific ICF environments.

phosphate, magnesium, amino acids, etc.—the ICF constituents.

Accumulation of phosphorylated metabolites (glucose 6-phosphate, etc.) in cells can also cause K movement into cells; this is only a minor contributor to the hypokalemia during therapy for diabetic ketoacidosis.

Cation-Cation Exchange

The exchange of K^+ for another cation can modify the ICF:ECF [K] ratio without changing the number of intracellular anions. This type of movement involves the NaKATPase, the enzyme dealing with cation transport across cell membranes. Cation movement also results in an ICF pH or volume change (see page 147).

THE NAKATPASE (see also page 215)

> THE NaKATPase IS RESPONSIBLE FOR RAISING THE ICF [K]. IT PUMPS 3 Na^+ OUT OF CELLS AND 2 K^+ INTO CELLS PER ATP USED.

This enzyme uses ATP to pump Na^+ out of cells and K^+ into cells. Its action results in a transmembrane [Na] gradient and the resting membrane potential (Fig. 9–2, page 215). Since the anions of the ICF are predominantly macromolecules that cannot cross the cell membrane, the net negative charge in the ICF is said to be "fixed." Although Na^+ or K^+ can balance these charges, two facts eliminate Na^+ from playing a major role in this regard. First, as soon as the [Na] in the ICF exceeds its normal value of close to 10 mM, it is actively pumped out of the cell by the NaKATPase (see also page 215). Second, the plasma membrane has a relatively low permeability to Na^+; hence Na^+ diffuse into most cells at a very slow rate under normal conditions.

HORMONES AND THE ICF [K] (Table 9–1, page 217)

> IMPORTANT HORMONES IN K SHIFTS:
> - INSULIN
> - β_2-ADRENERGICS
> - ALDOSTERONE (BASAL LEVELS)

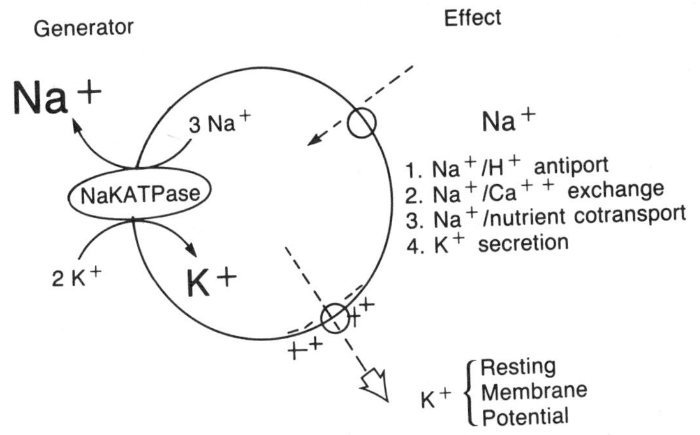

Figure 9–2. The role of the NaKATPase in ion transport–related activities. On the left-hand side of the figure, the action of the NaKATPase generates the [Na] and [K] gradients across the cell membrane. The effect of these gradients is shown on the right-hand side of the figure. The [Na] gradient drives the cotransport of materials with Na^+ (e.g., glucose and amino acid absorption in the GI tract and kidneys), the countertransport of Na^+ and H^+ for $NaHCO_3$ reabsorption and cell volume defense, Na^+-linked calcium movement, and renal K^+ secretion.

The [K] gradient leads to K diffusion out of cells and thereby generates the resting membrane potential.

NaKATPase Activity. There is normally a large excess of NaKATPase activity in the membranes of all cells and in the basolateral membrane of kidney and gastrointestinal cells. To achieve a greater rate of pump activity, the major requirement is a raised [Na^+] in the ICF. In the case of aldosterone in the kidney, a late action of this hormone is the induction of more NaKATPase enzyme units; this is accompanied by a large expansion of the basolateral membrane surface area. The general functions of the NaKATPase are shown in Figure 9–2.

The hormones discussed below have acute effects on K movement. Other effects such as those causing anabolism are long term and are not discussed here.

Insulin. Insulin causes K to shift into cells. Several mechanisms may be involved. For example, insulin causes an increase in the resting membrane potential, possibly owing to an increase in the NaKATPase activity, and this "attracts" K^+ into the ICF. Alternatively, insulin can cause the synthesis of phosphate esters in the ICF (hexose phosphates, RNA, etc.), and these anions "attract" K^+ into cells. Quantitatively, if the basal insulin concentration halves, the ECF [K] rises by about 0.5 mM within 30 min. Administration of insulin should cause a significant fall in the ECF [K]; this is especially important during treatment for diabetic ketoacidosis or treatment of hyperkalemia.

β-Adrenergic Action. Catecholamines play an important role in determining the distribution of K^+ across the cell membrane. Agonist-antagonist studies suggest that β_2 receptor agonists promote movement of K^+ into the ICF. In contrast, α-adrenergics tend to promote K movement out of cells during hypokalemia (see page 217 for more details).

Aldosterone. Although there are conflicting data, it appears that aldosterone causes K to move into the ICF. This action only seems to be quantitatively important during insulin deficiency. The combined deficiencies of insulin and aldosterone may lead to life-threatening hyperkalemia, especially if hyperglycemia is present.

As discussed in the section on K excretion, aldosterone has other and more striking actions on K homeostasis, owing to its action on the cortical collecting duct leading to K excretion in the urine.

Question (Answer on page 217)

7. With defense of cell volume, which ions must shift, and what are the consequences of these ion shifts?

Acid-Base Status and K Shifts Across the Cell Membrane

When HCl is given, there is a rise in the ECF $[H^+]$ and a

TABLE 9–1. Factors Causing a K Shift From the ICF to the ECF

Hormones
Insulin lack, β_2-adrenergic antagonists, aldosterone lack
Acid-Base Disturbances
Acute HCl gain or $NaHCO_3$ loss*
ICF Anion Change
Catabolism, loss of organic phosphates
Cell Necrosis
Rare Factors
Depolarization (e.g., succinylcholine)
Change in K^+ permeability (barium poisoning)
Unusual cation accumulations (lysine/arginine toxicity)
Hypertonicity (an overrated factor; see Answer 4 on page 211)

*Other acid-base disturbances have little net effect on K distribution.

Other Actions of Catecholamines on the Plasma [K]. The effects of catecholamines on the [K] in the ECF are complex.

β_2-Adrenergics stimulate K movement into cells.

β_1-Adrenergics stimulate renin release from the kidney, thereby increasing aldosterone production and K excretion.

Catecholamines stimulate glycogenolysis, which leads to hyperglycemia and insulin release; insulin causes K to move into the ICF.

α-Adrenergic action inhibits insulin release from pancreatic B cells; this tends to promote hyperkalemia; direct α actions lead to a K shift out of cells during hypokalemia.

Answer

7. A patient with hyponatremia must lose ICF compounds or ions to minimize cell swelling. Since the number of uncharged compounds is small, ions must be exported. The most abundant ICF cation is K^+; however, since the ICF [Cl] is not high in most cells, HCO_3 is required for ICF pH defense, and phosphates are critical for energy metabolism and vital functions (phospholipids, DNA, RNA, etc.), the ability to export K^+ is limited. Hence, only a minority of organs can defend their ICF volume, because when a cell extrudes K^+, the ECF [K] rises before K can be excreted by the kidney. Probably brain cells and possibly a few other tissues can defend their ICF volume in this manner (the brain is in a restricted space and swelling would be associated with compromised circulation and function).

Should the ICF lose Na^+ to defend cell volume, the ICF [Na] will fall. This will increase the tendency of Na^+ to move from the ECF to the ICF. Ions that countertransport with Na^+ (H^+, Ca^{++}) move in the opposite direction (see Fig. 9–2, page 215). Thus, defense of cell volume cannot be viewed as an event independent of defense of cell pH, Ca^{++}, and so on.

shift of H^+ into the ICF (close to 60% of the H^+ load is buffered in the ICF); K ions move out of cells for charge balance. In contrast to HCl, anions of the organic acids produced from carbohydrates (lactic acid) and fat (ketoacids) enter the ICF to almost the same degree as protons; hence, there is only a small K^+ shift as a result of ICF buffering of lactic acid and ketoacids (Fig. 9–3, page 219). Hence, hyperkalemia in metabolic acidosis due to the accumulation of these *organic acids* is the result of factors other than the acidemia per se.

During metabolic acidosis caused by the loss of $NaHCO_3$, H^+ are buffered in the ICF. Thus, this type of metabolic acidosis is associated with hyperkalemia early in its course. With time and a normal adrenal/renal axis, hyperkalemia should be corrected as a result of renal K excretion (Fig. 9–3, page 219).

- IN ACUTE METABOLIC ACIDOSIS, K MAY LEAVE THE ICF, AND DURING ACUTE METABOLIC ALKALOSIS, K ENTERS THE ICF
- ONLY METABOLIC ACIDOSIS CAUSED BY $NaHCO_3$ LOSS LEADS TO INITIAL HYPERKALEMIA OWING TO ICF BUFFERING. WITH TIME, A NORMAL KIDNEY AND ALDOSTERONE RESPONSE RETURNS THE PLASMA [K] TO NORMAL
- IN CONTRAST, A GAIN OF AN ORGANIC ACID DOES NOT CAUSE INITIAL HYPERKALEMIA: INSULIN LACK OR HYPOXIA MAY LEAD TO A K SHIFT INTO THE ECF AND TO HYPERKALEMIA
- K DOES NOT SHIFT WITH ACUTE RESPIRATORY DISORDERS

Respiratory acid-base disorders cause only small changes in the plasma [K]. Therefore, if the [K] changes with a P_{CO_2} change, look for a cause other than the simple acid-base disturbance (see page 221).

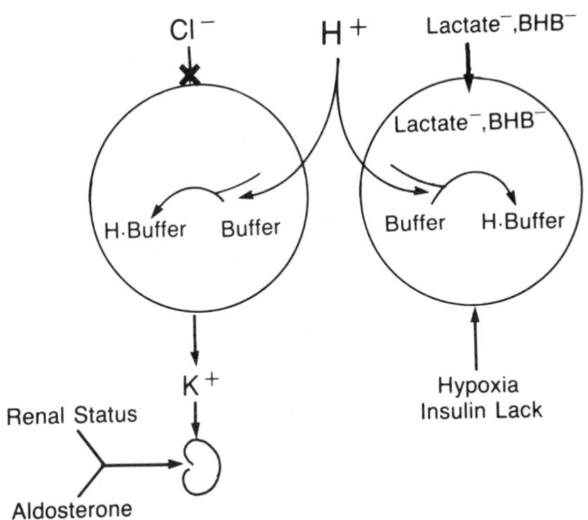

Figure 9–3. Buffering of H^+ and the consequent K^+ shift. The circles represent the cell membrane. Cl^- are restricted to the ECF, as depicted in the left-hand circle. When the $[H^+]$ in the ECF rises with a mineral acid load, H^+ enter the ICF and are buffered. In this situation, K^+ leaves the ICF for electroneutrality. In contrast, with organic acid loads, the organic anions (lactate⁻, β- hydroxybutyrate⁻) enter the ICF in close to an equal quantity to that of H^+; thus, little K^+ shift is required for electroneutrality (right-hand circle). In these situations, hyperkalemia is due to lack of oxygen (cell destruction) or lack of insulin.

Hyperkalemia should stimulate aldosterone release and result in a loss of this extra K in the urine. Therefore, a steady-state hyperkalemia in hyperchloremic metabolic acidosis should only be seen if there is renal insufficiency or low aldosterone bioactivity.

PHYSIOLOGY OF K RENAL EXCRETION

The Addition of K to the Luminal Fluid

Nephron Site. Almost all the filtered K^+ are reabsorbed prior to the distal nephron in normal kidneys. The K excreted in the urine is mainly due to K secretion in the late distal convoluted tubule and cortical collecting duct (abbreviated henceforth as the "cortical distal nephron").

Role of the Medullary Collecting Duct in K Excretion. Two major events occur in the medullary collecting duct with respect to the clinical interpretation of the excretion of K. First, water is reabsorbed here; this elevates the urine [K] (i.e., the K:H_2O ratio). This should be taken into account when assessing the urine [K]. The second event that occurs in the medullary collecting duct is the reabsorption or secretion of K. The magnitude of this movement is relatively small and only seems to be important for the urine [K] during hypokalemia or hyperkalemia.

- *VIRTUALLY ALL URINARY K IS ADDED IN THE "CORTICAL DISTAL NEPHRON"*
- ALDOSTERONE LEADS TO A HIGHER LUMEN [K] BY:
  - OPENING A Na^+ CHANNEL IN THE LUMINAL MEMBRANE
  - INCREASING THE ACTIVITY OF THE NaKATPase IN THE BASOLATERAL MEMBRANE.
  - OPENING LUMINAL K^+ CHANNELS
  - DECREASING LUMINAL Cl PERMEABILITY
- K EXCRETION MAY BE HIGHER WHEN THERE IS A LARGE VOLUME DELIVERED TO THE "CORTICAL DISTAL NEPHRON"

Mechanism of K Secretion. Aldosterone opens a luminal Na^+ channel. This has 2 effects: (1) it lowers the membrane potential across the luminal membrane, and this favors cation secretion into the lumen; (2) the higher [Na] in the cell increases flux through the basolateral NaKATPase, which brings K from the ECF into principal cells. The net effect is K movement from the ECF to the lumen. Said another way, aldosterone actions lead to an elevated luminal [K] that is approximately 10-fold higher than that in the ECF (Figs. 9–4 and 9–5, page 223).

K Shift in HCl vs Lactic- or Ketoacidosis. When 375 mmol of acid are given, 60% or 225 mmol of H^+ are buffered in the ICF. For this H^+ load to enter the ICF, close to 225 mmol of Na^+ plus K^+ must move in the opposite direction when the acid is HCl (roughly twice as much Na^+ as K^+ exits from the ICF during metabolic acidosis). In contrast, close to 225 mmol of anion (lactate, β-hydroxybutyrate) enter the ICF with the H^+ in the organic acidoses. Therefore, with a lactic acid or ketoacid load, there is a much smaller K shift from the ICF (Fig. 9–3, page 219).

K Shift in Acute Respiratory Acid-Base Disorders. During acute respiratory acidosis or alkalosis, there is little change in ECF [HCO_3] and only a very small K^+ movement across the cell membrane. In acute experiments, no appreciable change in ECF [K] was detected over a 4-fold change in $Paco_2$.

K Shift in Chronic Respiratory Acidosis. During chronic respiratory acidosis, the plasma [HCO_3] rises, but only a minor plasma [K] change should be anticipated. Three aspects of this HCO_3 rise are important with respect to "deducing" the magnitude of a possible K shift from the ICF.

a. The rise in [HCO_3] came about as a result of renal events that, in essence, were equivalent to the loss of NH_4Cl in the urine. In ECF terms, this is equivalent to the "exchange" of Cl for HCO_3 and is independent of H^+ and K^+ shifts across body cell membranes.

b. The magnitude of the HCO_3 gain in the ECF is less than the H^+ load shifted across the cell membrane in metabolic acidosis with an equivalent pH change. Hence, in chronic respiratory acidosis, a much smaller cation shift should be expected.

c. Adrenal/renal events would have occurred over this time period to minimize any rise in the plasma [K].

Clinical Analysis of Renal K Handling				
K excretion	=	Urine [K]	×	Urine volume
	=	Aldosterone action	×	Water load (hemodynamics)

Conclusion: To analyze renal K excretion in vivo, we need to assess both the urine [K] and urine volume.

Availability of Na^+ in the Lumen of the Distal Nephron. K cannot be secreted unless Na is delivered to the lumen of the distal nephron. The [Na] required for half-maximal rates of K secretion is 13 mM in the rat; hence, with a urine [Na] of 25 mM, K secretion is not limited by Na.

If a patient has a very low GFR, owing to low effective circulating volume, Na^+ delivery could limit this process.

Distal Nephron Flow Rate and K Excretion.

K EXCRETION = URINE [K] × VOLUME

Aldosterone raises the luminal [K] to almost 10-fold above the plasma [K] (Fig. 9–4, page 223). If the plasma [K] is 4 mM, then the luminal [K] is 40 mM (i.e., 10 × 4 mM). If distal delivery was 5 L, then 200 mmol of K could be excreted; alternatively, a 10-L delivery could result in the excretion of 400 mmol at the same transtubular [K] gradient. Hence, the importance of the volume of filtrate delivered to this nephron site can be appreciated.

The Adrenal Gland and Aldosterone Release

HYPERKALEMIA AND ECF VOLUME CONTRACTION ARE THE 2 MAJOR STIMULI FOR ALDOSTERONE RELEASE.

The release of aldosterone is stimulated by hyperkalemia, angiotensin II, or the aldosterone-releasing factor of the anterior pituitary gland (independent of ACTH). The first two stimuli act in a synergistic fashion. The level of angiotensin II rises when renin is released from the juxtaglomerular apparatus of the kidney. Renin release is stimulated by ECF volume contraction, renal artery stenosis, and β_1-adrenergic stimulation; in contrast, it is inhibited by ECF volume expansion and adrenergic β_1-blockers. Destruction of the juxtaglomerular apparatus by interstitial renal disease,

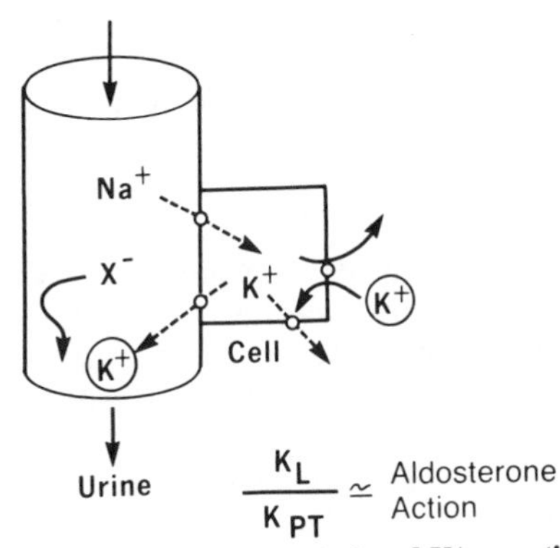

Figure 9–4. Physiology of aldosterone-induced K^+ secretion. The actions of aldosterone lead to an increase in the transtubular [K] gradient (see also Fig. 9–6). The barrel-shaped structure represents the lumen of the "cortical distal nephron" and the square a principal cell. The first major action of aldosterone is the movement of luminal Na^+ down the [Na] gradient through a Na^+ channel without its anion (X^-), which makes the lumen more negatively charged (aldosterone diminishes Cl^- conductance). Intracellular Na^+ are pumped out the basolateral membrane by the NaKATPase, an action that raises the intracellular [K]. ICF K^+ leak across both the luminal and basolateral membranes, more going into the lumen because of the electrical forces. The net result of all these events is K^+ addition to the lumen. The ratio of [K] in the lumen (K_L) to that in the peritubular space (K_{PT}) should provide an index of the K secretory process (usually aldosterone action).

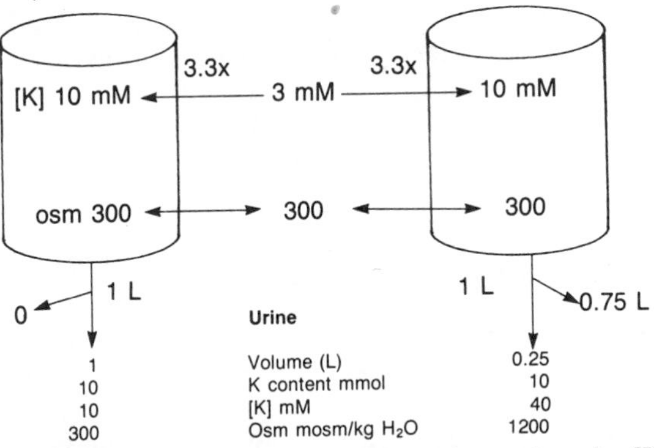

Figure 9–5. Role of medullary water reabsorption on the urine [K]. The barrel-shaped structures represent the cortical collecting duct and the arrows below them the medullary collecting duct. In both examples, there is a small transtubular [K] gradient; i.e., the luminal [K] is 10 mM or 3.3 × the peritubular [K] of 3 mM. For simplicity, consider what happens when 1 L of fluid traverses the medullary collecting duct. In the left-hand example, there is no medullary water reabsorption, whereas 75% of the water is reabsorbed in the right-hand example. In both cases, no K is reabsorbed

especially common in diabetes mellitus, also results in lower levels of renin release. In some cases, angiotensin II levels may not be elevated despite high renin levels. This can occur when drugs inhibit the conversion of angiotensin I to its active form, angiotensin II (converting enzyme inhibitors such as captopril). Atrial natriuretic factor diminishes aldosterone release from zona glomerulosa cells.

How to Assess Aldosterone Action in Vivo. A note of caution should be inserted because little direct data are available to verify the approach described below. Nevertheless, this approach should provide a rational basis for the clinical analysis of aldosterone action with respect to disorders of K excretion. The clinician should attempt to estimate the transtubular [K] gradient in the cortical distal nephron *in vivo* (TTKG). Expressed as an oversimplification, the [K] in the plasma should reflect that in the peritubular area of the cortex of the kidney (the errors introduced by this assumption will be small in most cases). The luminal [K] in the cortical distal nephron can be "back-calculated" from the [K] and osmolality of the urine (Fig. 9–5, page 223).

In summary, if one divides the urine [K] by the urine-to-plasma osmolality ratio

$$([K]_{urine}/(urine:plasma)_{osm})$$

a reasonable approximation of the [K] at the end of the cortical distal nephron can be deduced. This value will be high (7 or more times the plasma [K]) if mineralocorticoids are acting and low (less than 4 times the plasma [K]) if they are not. Be sure that the urine contains at least 25 mM Na and is <u>not hypotonic</u> before doing this analysis in order to reflect aldosterone action *in vivo*.

$$TTKG = [K]_{urine}/(urine/plasma)osm/K_{plasma}$$

Questions (Answers on page 227)

8. Can you determine whether a high urine [K] is due to K addition to the "cortical distal nephron" or to water reabsorption in the medullary collecting duct?
9. Does the excretion of K conserve Na and thereby defend the ECF volume? This question requires a sophisticated answer (think about where the K to be excreted comes from).

or secreted in the medulla. Therefore, although the K excretions are equal in both cases, the urine [K] is 4-fold higher (40 mM) when water is reabsorbed in the medulla (right side). This should be taken into account in assessing the urine [K].

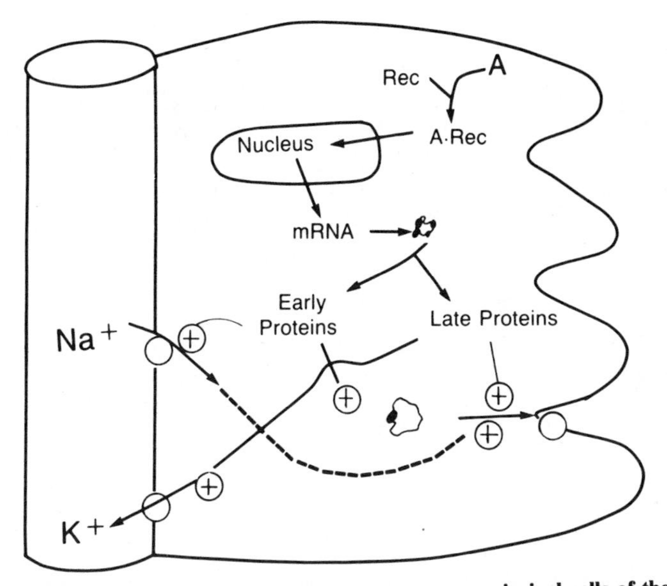

Figure 9–6. Summary of actions of aldosterone on principal cells of the "cortical distal nephron." Aldosterone (A) enters the principal cell via the basolateral membrane; once in the cytoplasm, aldosterone binds to a specific receptor (REC). This hormone receptor complex enters the nucleus and induces the synthesis of messenger RNA, which causes the synthesis of new proteins. For simplicity, we can divide these proteins into early and late effectors of aldosterone action. The early action is the opening of luminal Na^+ channels and, possibly, the synthesis of the precursor of the NaKATPase. The later actions of aldosterone include the insertion of more active NaKATPase units into the basolateral membrane and K channels into the luminal membrane. These later actions of aldosterone may be triggered by a high ICF [Na].

Suggested Reading

Adrogue HJ, Chap Z, Ishida T, et al: Role of the endocrine pancreas in the kalemic response to acute metabolic acidosis in conscious dogs. J Clin Invest 75:798–808, 1985.

Brown RS (discussant): Beth Israel Hospital and Harvard Medical School: Extrarenal potassium homeostasis. Kidney Int 30:116–127, 1986.

Field MJ and Giebisch GJ: Hormonal control of renal potassium excretion. Kidney Int 27:379–387, 1985.

Jamison RL, Work J, and Schafer JA: New pathways for potassium transport in the kidney. Am J Physiol 242:F297–F312, 1982.

Marver D and Kokko JP: Renal target sites and the mechanism of action of aldosterone. Miner Electrolyte Metab 9:1–18, 1983.

Sterns RH, Cox M, Feig PU, et al: Internal potassium balance and the control of the plasma potassium concentration. Medicine 60:339–354, 1981.

West ML, Bendz O, Chen CB, et al: Development of a test to evaluate the transtubular potassium concentration gradient in the cortical collecting duct in vivo. Miner Electrolyte Metab 12:226–233, 1986.

West ML, Marsden PA, Richardson RMA, et al: New clinical approach to evaluate disorders of potassium excretion. Miner Electrolyte Metab 12:234–238, 1986.

Wright FS: Sites and mechanisms of potassium transport along the renal tubule. Kidney Int 11:415–432, 1977.

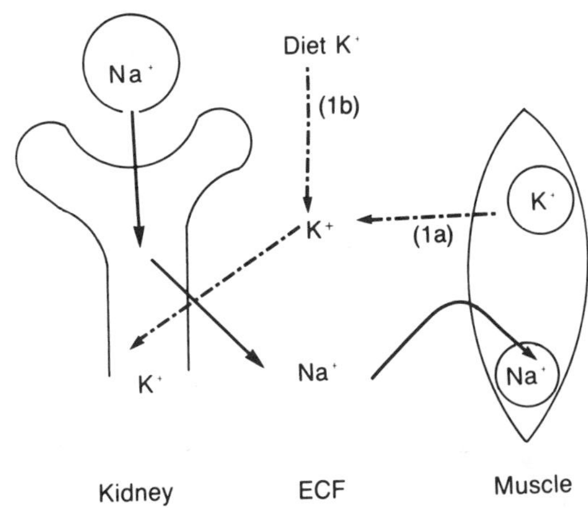

Figure 9–7. Source of excreted K determines if the ECF volume was defended. The reabsorption of filtered Na with K secretion will result in the preservation of the ECF volume only when the source of the K was dietary.

Answers

8. Yes. As described in Figure 9–5, page 223, a rise in the urine [K] due to water absorption in the medullary collecting duct is accompanied by a rise in the urine osmolality. Therefore, one should divide the urine [K] by the urine:plasma osmolality ratio to interpret the urine [K] with respect to cortical vs medullary events.

9. *Integration of the actions of aldosterone on Na and K physiology (Fig. 9–7):* The first step in aldosterone action is the reabsorption of Na in the cortical collecting duct. However, this reabsorption of filtered Na does not necessarily conserve Na for the ECF. Consider Pathway 1a for example. To reabsorb Na, K was transported from the ECF to the urine. Since the ECF has little K, the source of the K is really the ICF. To transport K^+ out of the ICF, Na^+ entered this compartment. Therefore, the net effect is the translocation of intracellular K^+ to the urine and ECF Na^+ to the ICF (Pathway 1a). These actions do not appear to defend either the ECF volume (Na content) or the ICF composition (K); however, in a broader context, they may, if (1) the source of the urinary K was dietary K (Pathway 1b), (2) Na reabsorption was linked to ammonium excretion (hypokalemia stimulates ammoniagenesis), or (3) NaCl was reabsorbed. Hence, it is necessary to obtain a broad overview to integrate Na, K, and acid-base physiology.

10

HYPOKALEMIA

THE BIG PICTURE

General Considerations. **Hypokalemia ([K] < 3.5 mM) only produces serious side effects when the [K] is < 3.0 mM. Since only 2% of K is in the ECF (60 mmol), total body K is monitored through a tiny and inadequate window. A plasma [K] of 3.0 mM occurs when close to 200 mmol of K are lost.**

Etiology of Hypokalemia. **Virtually all cases of chronic hypokalemia are due to increased K loss. K loss is usually caused by vomiting, nasogastric suction, diarrhea, or diuretics. The clues are found in the history (vomiting, laxative abuse, diuretics—all of which the patient may deny) and the physical finding of ECF volume contraction. If there is no ECF volume contraction, look for a primary reason for aldosterone excess. The presence of hypertension is a useful clue. A shift of K into cells can occur under the influence of hormones, metabolic alkalosis, or anabolism; however, these are rare causes for chronic *severe* hypokalemia.**

Diagnosis of Hypokalemia. **Clinical suspicion is**

K PHYSIOLOGY CAPSULE: CELLS

1. 98% of the K in the body is in cells.
2. The principal factor elevating the ICF [K] is the NaKATPase.
3. Passive diffusion of K^+ out of cells generates the resting membrane potential, and this is largely reflected by the transmembrane [K] gradient.

DAILY K BALANCE

1. Close to 1 mmol of K is consumed per kg of body weight/day (roughly equal to the ECF K content).
2. To excrete the daily K load in the urine, Na^+ plus fluid is delivered to the cortical distal nephron, where aldosterone acts to raise the luminal [K]; hormone action takes 1–2 hr. Therefore, an acute defense against hyperkalemia is needed—the temporary shift of K into cells.
3. Hormones (insulin, β-adrenergics, and aldosterone) lead to a K shift into cells. Acid-base factors are "overrated"; the accumulation of ICF anions (i.e., anabolism) can be important to shift K into cells.
4. Summary: Eat K, store it in cells temporarily, and then excrete it several hours later.

EXCRETION OF THE DAILY K LOAD IN THE URINE

Almost all filtered K is reabsorbed; the K in the urine was secreted in the cortical distal nephron.

K^+ secretion requires that Na^+ be reabsorbed; the Na^+ reabsorbed into the ICF of the cortical distal nephron is pumped out into the ECF via the NaKATPase.

Na reabsorption and the NaKATPase are both stimulated by aldosterone.

The [K] in principal cells rises and K^+ move passively into the luminal fluid.

In the presence of aldosterone, the luminal [K] is close to 10× that in the ECF.

Since aldosterone only elevates the luminal [K], factors that increase the urine flow rate also augment K excretion.

$$K_{excretion} = Urine_{[K]} \times \text{urine volume}$$

essential because weakness is a late symptom and is not present until K depletion is severe; hypokalemia may be found on "routine" electrolytes or as a result of ECG changes. Urine electrolyte levels are essential to establish the etiology. Interpret the urine [K] by adjusting the urine [K] for water reabsorption in the renal medulla.

Dangers of Hypokalemia. **The major danger is a cardiac arrhythmia, especially if digitalis and/or metabolic alkalosis is present. The danger of hypoventilation due to hypokalemia in metabolic acidosis may be aggravated by giving HCO_3 before K.**

Treatment. **Large quantities of K must be given when the total body deficit is great; however, a very large intravenous K bolus may cause arrhythmias. In most cases, KCl is needed; $KHCO_3$ should be given in severe metabolic acidosis. The patient receiving digitalis and the patient with diabetic ketoacidosis in whom K will shift into cells 1–2 hr following insulin administration must have K replaced more aggressively.**

ILLUSTRATIVE CASE

A patient with liver cirrhosis had a 5 L GI hemorrhage. He was given 11 units of washed, packed red blood cells and 5 L of isotonic saline. His condition stabilized, but he appeared unduly anxious (acute alcohol withdrawal). His plasma [K] declined from 3.8 to 2.8 mM; the blood pH rose from 7.40 on admission to 7.45, but the plasma [HCO_3] and [glucose] did not change appreciably over this time. Urine output was 3.5 L over the time period.

Questions (Answers on page 231)

1. How much K is present in 1 L of blood?
2. What is the total body K deficit?
3. Was a shift of K into cells a major cause of hypokalemia? Comment on the effect of alkalemia and hormones. Can transfused RBCs contribute to hypokalemia (by what mechanism and to what degree)?
4. What factors could have contributed to a renal loss of K? What tests might be helpful in this regard? What information do you need to determine whether aldosterone was acting on the kidney?

Answers

1. K in 1 L of Blood. RBCs have a [K] of 100 mM; 40% of the blood volume is RBC. Therefore, each L of blood has 40 mmol of K in RBCs and 2.4 mmol in the plasma phase (0.6 L × 4 mM).

2. Total K Deficit. If there was no major K shift, a 1 mM fall in the [K] signals a K deficit of 100–400 mmol.

If you did not transfuse any RBCs and blood loss was 5 L, another 212 mmol of K were lost (5 L × 42.4 mmol K/L; see Answer 1).

3. Shift of K. A K shift into cells rarely causes such a large change in the plasma [K]. Although respiratory alkalosis (alkalemia, but no rise in the plasma [HCO_3]) was present, it did not cause a significant fall in the ECF [K]. Of the hormones, insulin is not likely to be important because the blood sugar was low and did not change; β_2-adrenergic activity secondary to alcohol withdrawal and/or hypotension could have caused some K to shift into cells.

Transfused RBCs. Storage of RBCs in the cold causes K to leak out of these cells; washing away this K results in a K loss (up to 5 mmol of K/unit). Rewarming of these RBCs in the patient leads to an uptake of K. This would only make a small contribution to the overall K deficit.

4. Renal Loss of K. Aldosterone and a high urine flow rate cause K loss in the urine. ECF volume contraction leads to aldosterone release; reduced hepatic clearance of aldosterone due to cirrhosis of the liver might also play a role. Cortical distal nephron flow rate should rise consequent to ECF volume reexpansion. To confirm renal K loss, measure the urine volume and [K]. To test whether aldosterone is acting, assess the luminal [K] in the cortical distal nephron by adjusting the urine [K] for medullary water abstraction (divide by the urine:plasma osmolality ratio); if this adjusted [K] is 7-fold or more greater than the plasma [K], it is likely that aldosterone is acting on the kidney.

APPROACH TO PATIENTS WITH HYPOKALEMIA (see Flow Chart 10–A, page 233, and Fig. 10–2, page 239)

The basis for the approach is first to decide if the K deficit is due to renal K loss or not. If the K loss is renal, the major subgroups are those due to hyperaldosteronism and those associated with high distal nephron flow rates.

ETIOLOGY OF HYPOKALEMIA

HYPOKALEMIA IS ALMOST NEVER SOLELY DUE TO LOW K INTAKE

Hypokalemia Due to Decreased Intake of K. It is very rare to see K deficiency develop solely on the basis of low K intake, because renal excretion of K can decline to very low levels. It might take weeks to months to lower the body K content by 100 mmol (recall that the body contains 3000–4000 mmol of K; see page 223 for more details). However, a low K intake can contribute to hypokalemia if there is a source of K loss from the ECF. This becomes important when diuretics are given to people who eat a low quantity of K-containing foods (e.g., the elderly who are hypertensive).

There is another indirect relationship between K and diet. Certain materials such as resins bind K^+ and convert them to a form that is not absorbed. This can be considered a decrease in the "net K intake." Clinicians take advantage of this fact in therapy of hyperkalemia by attempting to promote gastrointestinal K loss with K-binding resins.

Increased K Entry Into Cells. The factors promoting K entry into cells are listed in Table 10–1, page 235 (see page 210 for details). In general, these factors, when acting individually, lead to a fall in serum [K] of less than 1 mM.

Role of Acid-Base Disorders

K SHIFTS INTO CELLS IN METABOLIC BUT NOT IN RESPIRATORY ALKALOSIS

K should shift from the ECF into cells as a result of H^+

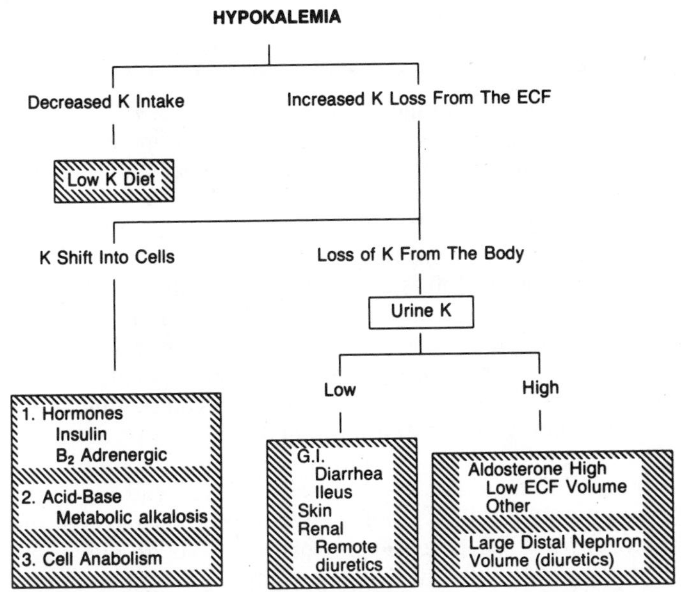

Flow Chart 10–A. Investigation of hypokalemia. The open box represents the parameter that, when evaluated, permits one to proceed to the diagnosis. The final diagnostic groups are shown in the shaded boxes (for a more detailed discussion, see the text).

Quantitative Aspects: A More In-Depth Look. The [K] gradient across the cell membrane is 150 mM/4 mM or 37.5. If the plasma [K] falls from 4 to 3 mM and this K gradient remains constant, the fall in ICF K content is 25% of the total or 800 mmol (0.25×3200 mmol). Since the total measured K deficit has been estimated to be only half of this value at best in clinical studies (100–400 mmol), the transcellular K gradient must rise (i.e., the ratio of [K] in the ICF to ECF is higher). A rise in this ratio implies that the resting membrane potential has risen (the cell membrane hyperpolarizes).

For the cell to lose 400 mmol of K, there must be a loss of 400 mEq of anions or a gain of 400 mmol of Na^+ and/or H^+. Hence, major changes in the ICF environment should be expected.

movement in the opposite direction (see Fig. 10–1, page 235, for details). Since the magnitude of this H^+ shift is much smaller with respiratory acid-base disorders, do not expect hypokalemia in patients with respiratory alkalosis; K shift into cells can occur during metabolic alkalosis.

Action of Hormones

> HORMONE-INDUCED K SHIFTS ARE DUE MAINLY TO INSULIN AND β_2-ADRENERGICS

INSULIN. Insulin promotes K entry into the ICF by increasing the resting membrane potential and by increasing the ICF anion content (phosphate esters). The major clinical setting in which this action is observed is following insulin administration to treat hyperglycemic disorders or possibly during endogenous insulin release stimulated by glucose ingestion. This K shift is transient and depends on the onset and duration of the action of the insulin administered. The change is usually relatively small in magnitude (0.5 mM), unless there is an underlying K deficiency as in diabetic ketoacidosis; in this case, the [K] change can be quite large (approximately 90 min after insulin administration) and requires immediate attention.

β-ADRENERGIC ACTIVITY. β_2-adrenergics lead to K entry into cells. Elevated levels of these hormones are seen during the response to hypotension, stress, and hypoglycemia, and in the clinical setting of delirium tremens. Drugs with these actions are commonly given to alleviate bronchospasm. With the usual dose of a β_2-agonist, the plasma [K] can fall by 0.8 mM, whereas massive overdoses cause a fall of almost 1.5–2.0 mM. In contrast, β_2-antagonists cause a much smaller rise in the plasma [K] (0.3 mM).

ALDOSTERONE. Aldosterone deficiency causing hyperkalemia is the only important action of aldosterone on cellular K distribution. Hypokalemia in patients with high aldosterone activity is due primarily to renal K loss rather than a K shift into cells.

Gain in ICF Anions. During growth, anionic compounds are synthesized (RNA, DNA, phospholipids, phosphoproteins, soluble organic phosphates); for electrical balance, K must also enter cells, and hypokalemia develops if K intake is not appropriately increased.

TABLE 10–1. Causes for a K Shift into Cells

Acid-Base Disorders
Metabolic alkalosis
Hormones
Insulin
β_2-Adrenergic agonists
Anabolism
Growth
Recovery from diabetic ketoacidosis
Total parenteral nutrition
Red cell synthesis (e.g., recovery from pernicious anemia)
Rare Disorders
Hereditary hypokalemic periodic paralysis
Other
? Anesthesia

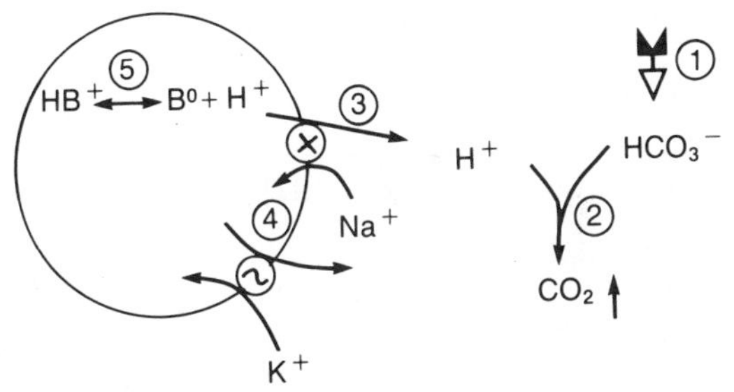

Figure 10–1. HCO_3 causes K to shift from the ECF to the ICF. The addition of HCO_3^- (1) causes the $[H^+]$ in the ECF to fall (2). This causes Na^+ to enter the ICF in conjunction with H^+ export on the Na^+/H^+ antiporter (3). The NaKATPase (4) pumps this ICF Na^+ out and pumps K^+ into the ICF. Continued flux requires back-titration of ICF buffers (5). In patients with a high plasma $[HCO_3^-]$ (metabolic alkalosis), look for additional reasons for hypokalemia (e.g., renal K loss secondary to diuretics or vomiting).

Other Causes for a K Shift Into the ICF. These examples are shown on page 237.

Hypokalemia Due to Enhanced K Loss

THE MAJOR SITE OF K LOSS IS RENAL. HYPOKALEMIA DUE TO ACTUAL GI K LOSS IS ALMOST ALWAYS CAUSED BY DIARRHEA

Gastrointestinal Tract. The [K] of upper gastrointestinal secretions is usually 15 mM or less. Since gastric secretions have such a low [K], hypokalemia due to vomiting or nasogastric suction actually results from K loss in the urine (see Answer 5, page 237, for more detail). In contrast, loss of colonic contents can lead directly to hypokalemia. In cholera, for example, 6 L of water containing 750 mmol of Na and 100 mmol of K can be lost. When diarrhea is due to more distal colon involvement (villous adenoma of the rectum), the K content might be 2–3-fold higher. Suspect that diarrhea is the cause of hypokalemia when metabolic acidosis with a normal plasma anion gap is present in the absence of renal K wasting.

Another type of gastrointestinal K loss occurs following the ingestion of materials (resins, clay) that bind K in the gastrointestinal tract. These causes are obvious from the history. Laboratory examination reveals a low urine [K] (see below for interpretation of urine [K] relative to medullary water abstraction).

Question (Answer on page 237)

5. What is the cause for hypokalemia in patients who vomit and have metabolic alkalosis? Is the K loss via the stomach important?

Urinary K Loss. Virtually all urinary K to be excreted is in the luminal fluid in the late distal convoluted tubule and the cortical collecting duct (for review, see Physiology Capsule II, page 229). Little may happen to this K during transit through the medullary collecting duct; however, water may be reabsorbed here and thereby increase the urine [K] (i.e., the K:H_2O ratio, Fig. 10–3, page 241). Thus, the major factors responsible for K *excretion* are the actions of aldosterone and the volume of fluid delivered to the cortical collecting duct (Na must also be delivered but, once delivered, does not directly regulate K secretion). Therefore,

Answer

5. *Vomiting is an example of K loss by more than one route.* As outlined in Figure 4–1, page 99, of Chapter 4, the plasma [HCO_3] rises and [Cl] falls with each loss of gastric contents. The consequent increase in the filtered load of HCO_3^- exceeds its renal reabsorption and leads to the loss of Na^+ from the body. The resultant decrease in ECF volume stimulates aldosterone release. Aldosterone action, together with a large distal nephron delivery of Na-rich fluid, produces a large renal loss of K. Therefore, vomiting leads to hypokalemia, owing to a large loss of K, but the route of this loss is primarily renal; to a lesser extent, there is a shift of K into the ICF (Fig. 10–1, page 235).

Gastrointestinal K losses per se are quite small, as the [K] of gastric fluid is usually less than 15 mM.

OTHER CAUSES OF A K SHIFT INTO THE ICF

Anesthesia. In animals, the induction of anesthesia leads to a significant fall in the plasma [K] that is independent of the acid-base status. It may be due to catecholamine release and/or a direct membrane effect of anesthetics. No comparable data are available in humans.

Hypokalemic Periodic Paralysis. This rare disorder is characterized by recurrent episodes of hypokalemia and weakness. The hypokalemia is aggravated by factors that normally cause a small shift of K into cells, e.g., carboydrate-rich meals (insulin), stress, or adrenaline administration.

The onset of hypokalemia is acute, and the degree of fall can be severe.

Treatment consists of avoiding precipitating factors, the use of β-blockers, and K administration. Long-term therapy may include K supplements and agents that diminish renal K losses (K-sparing diuretics). Acetazolamide is often effective, but the mechanism of its effect is unknown (despite the fact that this diuretic should increase renal K excretion, the plasma [K] rises). Finally, if thyrotoxicosis is present, it should be treated.

Hypothermia. Hypokalemia has been observed during hypothermia, and the converse is seen on rewarming.

we need only examine the causes for increased mineralocorticoids and cortical collecting duct volume delivery.

> CAUSES OF EXCESSIVE RENAL K LOSS:
> - HIGH ALDOSTERONE LEVELS
> - HIGH CORTICAL COLLECTING DUCT LUMINAL FLUID VOLUME

The causes for hyperaldosteronism are listed in Table 10–2, page 239. The clinician should look for two major groups: First, those with an enhanced stimulus for aldosterone release from a normal adrenal gland (in the absence of hyperkalemia, the major stimulus is an elevated renin concentration largely due to ECF volume contraction, a renin-producing tumor, or renal artery stenosis); second, there could be an increased autonomous aldosterone release due to stimulation by aldosterone-releasing factor, an adrenal tumor, or the presence of an agent with mineralocorticoid action. In this latter case, plasma renin levels should be very low.

Loss of K From Other Sources

SWEAT. Normally, K loss in perspiration is small ([K] tends to be 5–10 mM and the volume less than 1 L/day). However, in hot environments, more than 10 L of sweat can be produced per day. As a result of the Na loss and the ECF volume contraction, aldosterone increases and leads to enhanced skin and urine K loss.

LABORATORY INVESTIGATION FOR URINE K LOSS (Tables 10–2, page 239, and 10–3, page 241)

Urine [K]. Aldosterone causes the luminal fluid [K] to be 10-fold higher than that of the plasma (Fig. 10–3, page 241). Therefore, as hypokalemia becomes more severe, the luminal [K] must fall (it is now 10 × a lower plasma [K]). Furthermore, the rate of K excretion is directly proportional to the volume of fluid traversing this nephron site (if the luminal fluid [K] is 30 mM, 60 mmol of K would be excreted if 2 L flowed by; in contrast, K excretion would be 120 mmol if 4 L of fluid traversed this nephron segment). Hence, both the transtubular [K] gradient and the rate of K excretion should be evaluated. A urine [K] adjusted for water

TABLE 10–2. Causes of Hypermineralocorticoid States

Endogenous Aldosterone Release
- ***Associated With Increased Renin Levels***
  - ECF volume contraction*
  - Renal artery stenosis
  - Juxtaglomerular apparatus hypertrophy or tumor
  - Magnesium deficiency
  - Rare disorders such as Bartter's syndrome
- ***Associated With Low Renin Levels***
  - Primary adrenal cortical hyperplasia
  - Adrenal adenoma

Nonaldosterone-Mediated (associated with low renin levels)
- ***Endogenous Compounds With Mineralocorticoid Effects***
  - Cushing's syndrome
  - Adrenal gland enzyme defects
- ***Exogenous Compounds***
  - Administered steroids such as high doses of hydrocortisone
  - Licorice
  - Swallowed chewing tobacco
  - Amphotericin B (possibly)

*By far, the most common cause.

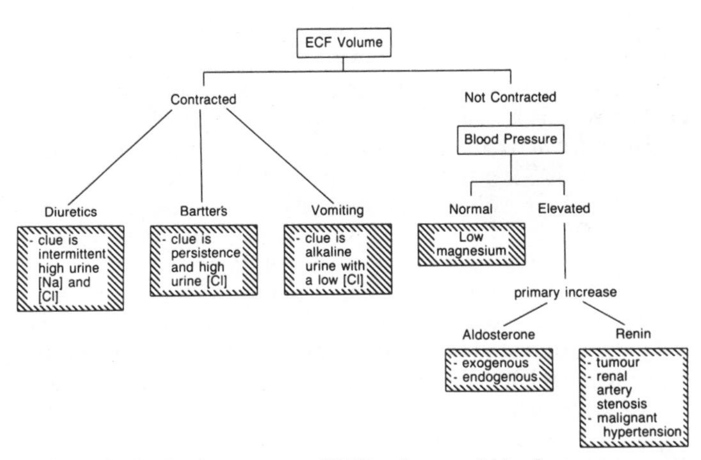

Figure 10–2. The importance of ECF volume and blood pressure assessments in the differential diagnosis of hypokalemia.

reabsorption that is less than 4-fold higher than that in the plasma suggests that mineralocorticoids are not acting on the kidney.

24-Hour Urine K Excretion. In a hypokalemic patient, a renal cause for K loss is present if the 24-hr urine contains more than 25–30 mmol of K. This test is more sensitive if the ECF volume is reexpanded while the patient is still hypokalemic. However, this test only documents that renal K wasting is a cause for the hypokalemia; it does not help define the mechanisms responsible for this K loss.

Importance of the Urine Anions in the Differential Diagnosis of Hypokalemia (Table 10–3, page 241). The two common causes for renal K wasting are vomiting and diuretics. In each case, ECF volume contraction due to Na loss stimulates aldosterone release and action.

Urine HCO_3. The cause for Na loss with vomiting is the excretion of HCO_3; more HCO_3 is filtered than the kidney can reabsorb—hence, filtrate is "dragged" to the cortical distal nephron, where aldosterone acts (to reabsorb Na^+ and secrete K^+). In vomiting, the patient reabsorbs all the NaCl possible so that almost no Cl is present in the urine.

Urine Cl. When a patient takes a diuretic, which inhibits NaCl reabsorption in the loop of Henle, more NaCl is delivered to the distal nephron than this region can reabsorb. Thus, Na deficiency, ECF volume contraction, and aldosterone release cause renal K wasting. This urine usually contains NaCl plus KCl. There is no excessive filtered load of HCO_3, so this anion does not appear in the urine.

Other Urine Anions. Should a patient have hypokalemia due to renal K wasting with a urine [Cl] close to zero mM (owing to ECF volume contraction) and the urine does not contain HCO_3, the clinician must search for the "missing urinary anions." The leading possibilities would include ketone body anions or anions from drug ingestion (e.g., acetylsalicylate, carbenicillinate, to name just 2). In this case, the sum of [Na] plus [K] greatly exceeds the sum of [Cl] plus [HCO_3], indicating that other anions are present in the urine. This clue can help the clinician understand the pathophysiology of renal K wasting and to design the appropriate therapy.

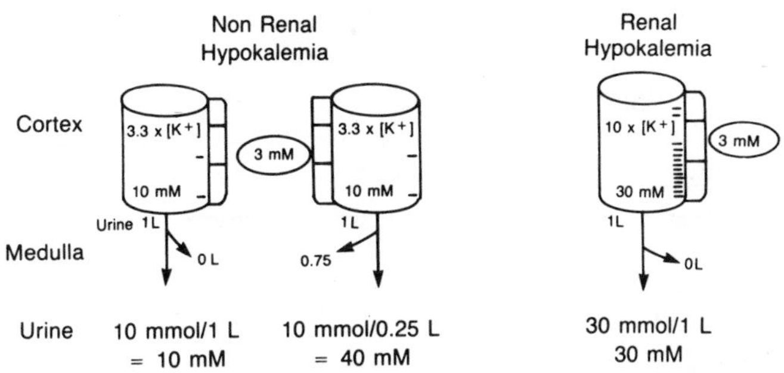

Figure 10–3. Factors responsible for elevating the urine [K]. The barrel-shaped structure represents the cortical collecting duct; the arrow below this structure represents the medullary collecting duct. The two left panels demonstrate the effect of water abstraction in the medulla on the urine [K]. In this example, the luminal [K] is 3.3 times the peritubular [K] of 3 mM or 10 mM. In Panel A, no water is abstracted in the medullary collecting duct, so the final urine [K] is also 10 mM; in Panel B, 3/4 of the water but no K is absorbed in the medullary collecting duct—thus the urine K/H_2O is increased 4-fold without changing the K content or K excretion rate.

In contrast, the luminal [K] is 10 times higher than the peritubular value of 3 mM (30 mM) in Panel C, reflecting aldosterone action. If no water is absorbed in the medulla, the final urine [K] is lower than that in panel B. However, adjusting for water abstraction by dividing by the urine/plasma osmolality ratio allows back-calculation of the luminal [K] in the cortical "distal nephron." Dividing this value by the plasma [K] should reflect the K secretory process.

TABLE 10–3. Urine Electrolyte Levels During Chronic Severe Hypokalemia of Renal Origin

Electrolyte	Vomiting	Diuretics
Na	• High if recent vomiting despite ECF volume contraction; otherwise <10 mM	• Less than 10 mM unless recent diuretic use
K	• High relative to blood value; absolute value depends on medullary water abstraction • Urine [K] ≥7 × plasma [K]	• Less than 10 mM unless recent diuretic use
Cl	• Always close to zero	• Less than 10 mM unless recent diuretic use
HCO_3	• Abundant with recent vomiting	• Zero

Questions (Answers on page 243)

6. What is the cause for K loss in patients A, B, C, and D? They all have EF volume contraction and metabolic alkalosis.

	Patient			
	A	**B**	**C**	**D**
Urine Electrolyte				
Na (mM)	25	25	25	0
K (mM)	60	60	60	10
Cl (mM)	0	85	0	0
HCO_3 (mM)	85	0	0	0

7. Will a patient consuming a very high NaCl load become hypokalemic owing to enhanced renal K excretion? How is this patient different from one receiving a loop diuretic, with respect to K homeostasis? (Assume equal Na excretion rates.)

Plasma Aldosterone and Renin. In the patient with excessive renal K loss, hypertension, and a normal ECF volume, the plasma renin and aldosterone levels help to differentiate adrenal from nonadrenal causes of hyperaldosteronism (Fig. 10–2, page 239). The diagnostic efficacy of these hormone assays is improved by expanding the ECF volume with saline to suppress endogenous renin release; these measurements should be made while the patient is recumbent to avoid a stimulus to renin release by ECF volume redistribution in the upright position.

Factors Influencing Flow Rate in the Distal Nephron. Although an increase in flow rate in the distal nephron augments K excretion, aldosterone action in conjunction with this increased flow rate leads to even higher rates of K excretion.

There are two major settings associated with high delivery volumes to the cortical collecting duct:

a. Those in which there is a marked increase in Na intake (especially if the anion accompanying Na is not Cl, e.g., HCO_3, drugs, etc.), causing more Na to be filtered than was reabsorbed;

Answers

6A. The major clue is the anion present in the urine. In this case, it is HCO_3. Since the patient has high HCO_3 content (metabolic alkalosis) and is losing HCO_3 in the urine, there must be an additional source of HCO_3. In the absence of HCO_3 intake, vomiting is the most likely cause.

6B. The major clue is the high urine [Cl]. Loop diuretics cause the excretion of NaCl, ECF volume contraction, and aldosterone release, which leads to a high [K] in the urine. A high urine [Na] (and [Cl]) in the face of ECF volume contraction suggests that the diuretic is acting on the kidney at the time the urine was collected. Alternatively, the patient could have Bartter's syndrome, if no diuretics were taken.

6C. The major clue here is that the sum of measured cations (Na^+ and K^+) greatly exceeds that of the usual anions (Cl^- and HCO_3^-). Therefore, another anion such as ketone bodies or drugs is excreted initially as their Na salts. This leads to ECF volume contraction, aldosterone release, and K loss in the urine.

6D. There is no excessive renal K loss at present; the kidney is behaving normally by conserving Na and Cl. If the delivery of Na and volume to the lumen of the distal nephron is low, there is little for aldosterone to act on. Thus, the diagnostic features of urine electrolytes disappear when the provocative stimulus (vomiting/diuretics) is not a recent event. You can suspect their presence in the past if treatment with NaCl plus KCl readily returns the plasma electrolyte values to normal. Furthermore, frequent "spot" urine electrolyte studies should be done to see if a sudden rise in Na, K, Cl and/or HCO_3 diuresis ensues. If it does, either the patient vomited (urine [HCO_3] high) or took diuretics (urine [Cl] high). In the latter case, do appropriate urine studies to measure the diuretic *at this time* if you wish to confirm your suspicion of occult diuretic abuse.

7. There must be a stimulus for aldosterone release plus a high distal flow rate to get very high rates of K excretion. This patient has a high intake of NaCl that increases distal flow rate but does not have ECF volume contraction to stimulate aldosterone release; hence, the patient will not become hypokalemic. Thus, he differs from the patient taking a diuretic that caused ECF volume contraction and aldosterone release in the setting in which the diuretic increases the distal flow rate.

b. Those in which there is reduced reabsorption of filtrate before the cortical distal nephron. The causes for reduced Na reabsorption at earlier nephron sites are the presence of diuretics (pharmacologic and osmotic) and the relatively uncommon primary diseases such as proximal renal tubular acidosis (especially if treated with $NaHCO_3$) and possibly Bartter's syndrome. In contrast, high flow rates in the absence of ADH action do not result in high K excretion rates (ADH is required for transcellular K transport in the cortical distal nephron).

Other Factors. Patients with Mg depletion have high aldosterone levels despite hypokalemia. The mechanism for aldosterone release is not clear. Suspect Mg depletion if hypocalcemia, alcoholism and/or a GI problem are present or if the patient was treated with drugs causing Mg wasting (loop diuretics, cisplatin, or aminoglycosides).

SYMPTOMS OF HYPOKALEMIA

Unless the plasma [K] is less than 3 mM, few symptoms will be present. These symptoms have a high degree of variability from patient to patient. A classification of these symptoms with respect to organ of origin is shown in Table 10–4, page 245.

Skeletal Muscle. The most common complaint related to hypokalemia is weakness or easy fatigability. Symptoms become prominent at plasma [K] below 3.0 mM and can ultimately lead to paralysis. This is especially important during metabolic acidosis because weakness can compromise ventilation. The pattern of muscle weakness due to K deficiency seems to be most evident in the lower extremities. With time and/or continuing K loss, the trunk and upper extremities become involved. Rarely, patients may also complain of cramps, muscle tenderness, and tetany. With severe hypokalemia, the patient is at greater risk of developing rhabdomyolysis. Should the hypokalemia go untreated for a prolonged time, muscle atrophy could occur.

Cardiac Muscle. Although controversy exists, it is likely that a number of cardiac arrhythmias may develop in severely hypokalemic patients. This is especially true if the patient is on digitalis or has underlying myocardial disease (see page 245).

Renal Problems. Polyuria and nocturia may develop during severe hypokalemia, owing to a reduction in renal

TABLE 10–4. Symptoms of Hypokalemia

Skeletal Muscle	
Weakness	
Rhabdomyolysis with extreme K depletion	
Cramps, myalgias	
Smooth Muscle	
Intestine	
Constipation, ileus	
Cardiac Muscle	
Arrhythmias (especially if digitalis or heart disease is present)	
Renal	
Concentrating defect (polyuria, nocturia)	
Enhanced ammonia synthesis (asymptomatic, unless hepatic insufficiency present)	
Neurologic	
Thirst	
Decreased deep tendon reflexes	
Paresthesias	

Controversy Concerning Hypokalemia and K Therapy. The clinician must decide whether a hypokalemic patient needs to receive K supplements. Apart from the replacement of existing deficits (to avoid more serious degrees of K depletion), the basis of K therapy is to prevent serious cardiac arrhythmias. On the opposite side of the argument, there is a significant cost: the real danger of hyperkalemia in patients on certain drugs (K-sparing diuretics, β-blockers), or with poor renal function. There is the lack of proof that treating modest hypokalemia (3.0–3.5 mM) really does diminish the cardiac arrhythmias and improve the patient's well-being. If the patient is taking a digitalis preparation, KCl treatment should be given. Furthermore, in other acute clinical situations in which there are factors causing a K shift out of cells that will be remedied shortly (treatment of severe metabolic acidosis), K therapy for hypokalemia is mandatory. Therefore, at present, prudent patient selection should avoid most of the dangers of KCl therapy.

ECG Changes During Hypokalemia

Hypokalemia tends to increase the resting membrane potential. Nevertheless, although ECG changes occur, they correlate poorly with the plasma [K]. In general, the earlier changes include T-wave flattening or even inversion. The S-T segment becomes depressed and may have a "trough-like" appearance. A "U" wave may also appear. In addition, severe hypokalemia tends to promote arrhythmias and conduction disturbances. The Q-T interval is normal or may be prolonged during hypokalemia, and this may help differentiate the ECG effects of hypokalemia from those seen with digitalis toxicity. The ECG abnormalities of hypokalemia are exaggerated if hypercalcemia is also present.

concentrating ability. These symptoms occur when the K deficiency is chronic and severe (greater than 200 mmol). The renal lesion may ultimately become irreversible, owing to interstitial fibrosis and tubular damage. In contrast, the ability to excrete a dilute urine is relatively preserved during hypokalemia. Polydipsia is also a common complaint during severe hypokalemia. This may be secondary to the polyuria or a more direct effect on the thirst center.

Other Factors. The central nervous system seems to be spared involvement in hypokalemia. The CSF [K] is constant despite wide swings in the plasma [K].

The ECG changes associated with hypokalemia are summarized on page 245.

TREATMENT OF HYPOKALEMIA

Since hypokalemia may have many causes, there is no universal therapy. We will only consider the replacement of the K deficit in this section. The urgency for this replacement must be dictated by the presence of digitalis, the possibility that K will shift into cells (recovery from diabetic ketoacidosis, β_2-adrenergic administration), heart disease, severe weakness, plus the severity of the deficit together with an assessment of continuing K losses. Monitoring of the ECG as well as assessing the possibility of hypoventilation and the degree of weakness are helpful.

When considering the quantity of K to administer, the aim should be to get the patient out of danger quickly but to replace the total K deficit more slowly. How quickly can K be administered? The consideration here is to avoid delivering a large bolus of K to the heart because this can be fatal. Hence, proper mixing of K in the venous blood is essential (given through a large but not central vein).

K Deficit

It is difficult to relate accurately the degree of hypokalemia to the extent of total body K deficit. A fall in the plasma [K] from 4 to 3 mM is associated with a total body K deficit of 100–400 mmol; a much larger deficit is required to lower the plasma [K] from 3 to 2 mM. However, if the patient also has a K shift into or out of cells, the magnitude of this deficiency is not reflected by these numbers.

TABLE 10–5. Indications for Initiating K Therapy During Hypokalemia

Absolute Indications
Digitalis therapy
Therapy for diabetic ketoacidosis
Symptomatic (e.g., respiratory muscle weakness causing hypoventilation)
Severe hypokalemia ($<$3.0 mM)
Strong Indications
Myocardial disease
Anticipated hepatic encephalopathy
Anticipated increase in another factor causing a K shift into the ICF (e.g., β_2-adrenergics)
Modest Indications
Development of glucose intolerance
Mild hypokalemia ([K] closer to 3.5 mM)
Need for better antihypertensive control

TO CORRECT SEVERE HYPOKALEMIA, SEVERAL HUNDRED mmol OF K ARE REQUIRED

Route of K Administration. The safest route to give K is by mouth. The conditions demanding intravenous therapy are gastrointestinal problems limiting K intake or absorption, severe hypokalemia with either respiratory muscle weakness or cardiac arrhythmias (especially if digitalis is present), and an anticipated K shift into the ICF (recovery from diabetic ketoacidosis) (Table 10–5, page 247).

Rate of Administration. The rate of K infusion should not exceed 40–60 mmol/hr in order to avoid delivering too large a K bolus to the heart. The [K] in the intravenous solution should be less than 60 mM for use in peripheral veins because higher [K] lead to local discomfort, venous spasm, or sclerosis. In general, do not administer K in glucose-containing solutions because the glucose itself (via insulin) can lead to an initial lowering of the plasma [K].

DO NOT GIVE K AT GREATER THAN 60 mM OR FASTER THAN 60 mmol/hr

Question (Answer on page 248)

8. What is the time course and the expected degree of hypokalemia with diuretic therapy? At the usual doses, is the hypokalemia particularly more severe with one class of diuretics?

Preparation of K. The anions accompanying K fall into three major classes. The choice depends on the clinical situation:

KCl. Cl is essential to correct the K deficit in all cases in which hypokalemia is associated with ECF volume contraction. Diuretics and vomiting (the 2 most common clinical causes for K deficiency) are associated with a Cl and ECF volume deficit. Cl permits Na to be reabsorbed in the proximal nephron, which diminishes the volume of filtrate delivered to the cortical distal nephron. This restores the ECF volume (suppresses renin and aldosterone release). Finally, Cl seems to be required for K reabsorption in the early distal convoluted tubule.

Answer

8. The plasma [K] falls to its steady-state value within one week of diuretic therapy. The degree of hypokalemia is modest—only 5% of patients have values less than 3.0 mM (if lower than 3.0 mM, look for another cause for K loss, including reasons for high aldosterone and/or urine flow rate). Female caucasians seem to be the most susceptible to developing hypokalemia.

The degree of hypokalemia may be slightly greater with thiazides than with loop diuretics. However, this is, at best, a very modest difference.

There are two factors that tend to increase the rate of K excretion: higher urine flow rates and higher urine [K]. The former will be elevated in patients taking both diuretics and a high salt load; the latter will be elevated by ECF volume contraction. Patients with a low Na intake have higher aldosterone levels, which promote greater K loss when diuretics are given.

KCl can be obtained as a liquid or in solid form. Enteric-coated KCl can lead to the development of small intestinal ulcerations due to local irritation by hypertonic KCl. The liquid form of KCl is relatively unpalatable; thus, KCl in a wax matrix or coated microspheres may be preferable.

***$KHCO_3$*.** Ingesting K salts of organic acids (acetate, citrate, gluconate, etc.) is equivalent to ingesting $KHCO_3$ because HCO_3 is produced when these organic anions are metabolized. Administer these salts when the patient has lost $KHCO_3$, i.e., diarrhea, etc. Obviously, this is a poor choice for K replacement if the patient has a high plasma [HCO_3] (metabolic alkalosis) or if the patient is unlikely to metabolize the accompanying anion (hypoxia, etc.).

> GIVE $KHCO_3$ ONLY IF THE PATIENT *NEEDS* THE HCO_3 LOAD

***K-Phosphate*.** If the K deficit is due to or associated with a loss of intracellular anions (phosphate), the K deficit is repaired only if phosphate is administered. Clinically, the need for K phosphate is most evident in ICF resynthesis during TPN or the recovery from diabetic ketoacidosis. However, clinicians may satisfy this need indirectly by giving KCl, while relying on the patient to eat the phosphate (plus Mg, etc.) required for ICF replacement. Recall that a large phosphate load has the danger of inducing hypocalcemia and metastatic calcification.

> GIVE PHOSPHATE (< 6 mmol/hr) IF YOU WANT K TO STAY IN THE ICF DURING ANABOLISM

***Dietary K*.** This is the best way to replace K deficits or keep up with ongoing losses. The diet should contain the phosphate, Cl, magnesium, and HCO_3 needed. Foods rich in K are fresh fruits, juices, and meat. However, a disadvantage may be that these foods contain a significant number of calories—a problem for obese patients.

Suggested Reading

Adrogue HJ and Madias NE: Changes in plasma potassium concentration during acute acid-base disturbances. Am J Med 71:456–457, 1981.

Artega E, Klein R, and Biglieri EG: Use of the saline infusion test to diagnose the cause of primary aldosteronism. Am J Med 75:722–728, 1985.

Haber E: The renin-angiotensin system and hypertension. Kidney Int 15:427–444, 1979.

Stein JH: Pathogenetic spectrum of Bartter's syndrome. Kidney Int 28:85–93, 1985.

Stuart LL, Peterson LN, Anderson RJ, et al: Mechanism of renal potassium conservation in the rat. Kidney Int 15:601–611, 1979.

Tannen R: Diuretic-induced hypokalemia. Kidney Int 28:988–1000, 1985.

11

HYPERKALEMIA

THE BIG PICTURE

General Considerations. **Hyperkalemia may produce serious side effects when the plasma [K] is >6 mM. The correlation between the plasma [K] and untoward effects (arrhythmias) is not reliable; in general, acute hyperkalemia produces more problems for any given plasma [K].**

Etiology of Hyperkalemia. **Chronic hyperkalemia is caused by a low rate of K excretion in almost every case. In the absence of renal failure, suspect low aldosterone bioactivity or diminished distal nephron flow. A shift of K out of cells is a rare cause for *chronic severe* hypokalemia; it occurs with hormone deficiency, metabolic acidosis, or catabolism.**

Diagnosis of Hyperkalemia. **Hyperkalemia is suspected clinically and confirmed with a plasma [K]. Urine electrolytes help establish the etiology.**

Treatment. **Eliminate K intake and promote K loss in the urine if hypoaldosteronism or a low urine flow**

TABLE 11–1. Approach to a Patient With Hyperkalemia

Rule Out Pseudohyperkalemia
Blood cell lysis
- Red blood cells
- Fragile tumor cells in blood
- Thrombocytosis

Assess Increased K Intake
May aggravate hyperkalemia if there is compromised renal K loss

Assess a Shift of K From the ICF to the ECF
Metabolic acidosis (nonorganic)
β_2-adrenergic blockade
Insulin deficiency
More than one hormone deficiency
Necrosis or depolarization
Rare disorders (e.g., hyperkalemic periodic paralysis)

Assess the Cause for Reduced K Loss in the Urine
Renal failure
Low aldosterone bioactivity
Low distal nephron flow rate

Therapy for Hyperkalemia
No K intake
Shift K into cells
Promote K loss in the urine, GI tract, or via dialysis
Antagonize effects of hyperkalemia

rate is present. Shift K into cells with insulin and/or bicarbonate (1–2 hours). Induce K loss in the GI tract with resins (slow); antagonize the cardiac effects of K with calcium (minutes); be prepared to promote K loss by dialysis if severe or rapidly progressive hyperkalemia is present.

ILLUSTRATIVE CASE

A 52 year old noninsulin-dependent diabetic is currently in poor glycemic control (blood glucose 360 mg/dl, 20 mM). His plasma [K] is abnormally high (6 mM). On physical examination, there is evidence of ECF volume contraction. Pertinent laboratory values reveal a modest metabolic acidosis (pH = 7.33, $[HCO_3]$ = 18 mM), no elevation in ketone bodies, a reduced GFR (serum creatinine = 2 mg/dl, 180 uM), a urine volume that is high with the following electrolyte concentrations: Na = 40 mM, K = 20 mM, Cl = 70 mM, pH = 5.4, osmolality = 600 mOsm/kg H_2O.

Questions (Answers on page 255)

1. What factors contributed to hyperkalemia?
2. What additional information is needed to understand the relative importance of each factor?
3. What therapy would you recommend?

APPROACH TO THE PATIENT WITH HYPERKALEMIA (Table 11–1, page 253, and Flow Chart 11–A, page 256)

The first thing to do is be certain that the high plasma [K] is not due to an artifact (called "pseudohyperkalemia"). The most common cause for this error is lysis of blood cells. If red cells have lysed, the plasma is pink to red in color. Rarely circulating tumor cells (leukemias) can be unduly fragile. Platelets can rupture if they contact glass or during clotting; however, significant hyperkalemia is only evident if the platelets are abnormally large and abundant in number (thrombocytosis). To minimize these errors, blood should be drawn gently into a heparinized silicone-coated tube to avoid lysis caused by a glass surface and during clotting.

Answers

1. First, pseudohyperkalemia due to a hemolyzed sample, thrombocytosis, or leukemia was ruled out. K intake was not excessive; therefore, the hyperkalemia was due to a K shift from the ICF and/or reduced K excretion.

K Shift from the ICF. Since necrosis is absent, suspect hormone deficiency. Insulin deficiency is present; aldosterone deficiency is discussed below. In the setting of insulin plus aldosterone deficiencies, hyperglycemia exacerbates hyperkalemia. Since metabolic acidosis was not ketoacidosis, it may have caused a significant K shift out of cells.

Reduced K Excretion in the Urine. Since the GFR is not markedly reduced and the urine volume is large, aldosterone deficiency is a good bet to explain the low rate of K excretion. Note that the patient has ECF volume contraction but is losing Na in the urine. Furthermore, the urine [K] is low (especially once you correct for water abstraction in the medullary collecting duct); thus transtubular [K] gradient is in the range suggesting low aldosterone bioactivity; see page 233).

2. The patient received insulin (blood glucose concentration fell to less than 180 mg % [10 mM]); however, the plasma [K] fell only to 5.8 mM.

The patient received sufficient $NaHCO_3$ to bring the plasma [HCO_3] to normal; the plasma [K] fell to 5.3 mM; however, the Na load could have led to an increased renal K excretion.

Bloods were drawn for aldosterone and renin. Since the results will not be back for several days, a physiologic dose of mineralocorticoids was administered. There was little change in the plasma and urine [K]. Thus, the major problem seems to be reduced K excretion in the presence of normal concentrations of mineralocorticoids. Very large doses of mineralocorticoids caused normokalemia and appropriately high rates of K excretion. Hence, there is a problem in the kidney (the plasma renin and aldosterone levels were subsequently found to be high). Thus, the final diagnosis is end-organ (kidney) hyporesponsiveness to aldosterone, possibly related to diabetes mellitus.

3. Therapy consists of a low K intake, insulin for better glycemic control (and to shift K into the ICF), $NaHCO_3$ to correct the metabolic acidosis (and to shift K into the ICF), and reassessment for longer-term mineralocorticoid requirement. Na is given to reexpand the ECF volume. Since end-organ hyporesponsiveness to aldosterone may be transient, the patient needs to be observed closely.

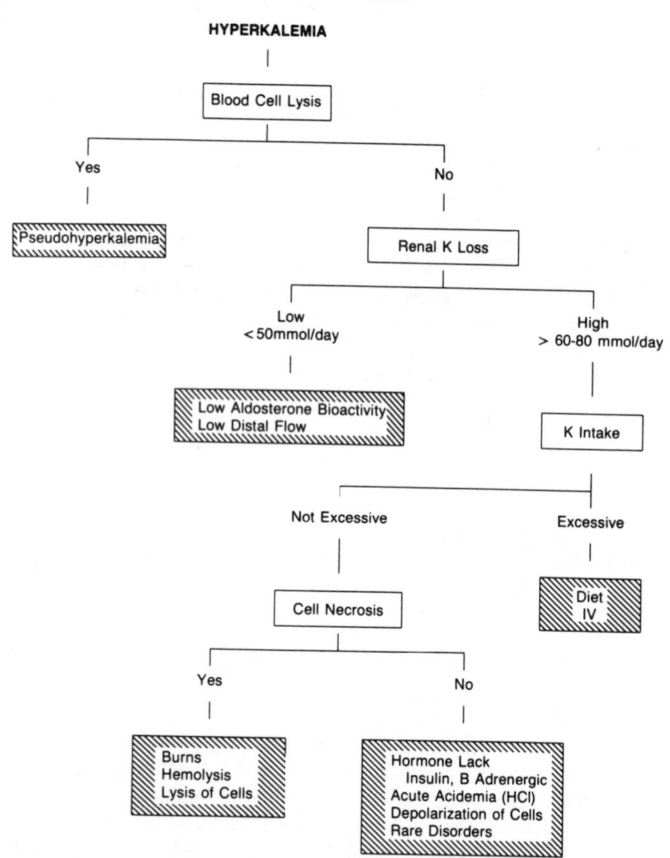

Flow Chart 11–A. Approach to hyperkalemia. The open boxes represent the variables that, when evaluated, permit one to proceed to the diagnosis. The final diagnostic group is shown in the shaded box.

Questions (Answers on page 257)

4. How much K is there in the platelets in 1 L of blood? (Assume 400,000 cells/mm^3, an ICF [K] of 150 mM, and a platelet volume 1 u^3.)

5. How much does the plasma [K] change if 1 L of pure water shifts from the ICF to the ECF? (Assume no change in the resting membrane potential and weight, 70 kg.)

6. If all the K ingested in one day (70 mmol) were retained without a change in resting membrane potential, how much would the plasma [K] rise?

Answers

4. The platelet count in normal blood is 4×10^5 cells/mm^3. This is equivalent to 4×10^{11} cells/L (1 cm^3 = 1 ml).

If the volume of an individual platelet is 1 u^3 (equal to 1×10^{-15} L), the total platelet ICF volume is the product of these two numbers, or 4×10^{-4} L.

Since the ICF [K] of a platelet was given as 150 mM, the total K content of platelets is 150 mmol $\times$ 4 $\times$ 10^{-4} L, or 6×10^{-2} mmol. If lysed, the rise in the [K] would only be 0.06 mM.

To get an appreciable rise in the serum [K] from the platelet disruption, the platelets must be increased considerably in volume; a change in the platelet number is much less important.

5. The [K] ratio across the cell membrane remains constant at 37.5:1 (constant resting membrane potential). When 1 L of ICF water is lost, the ICF volume declines from 28 to 27 L. Therefore, the ICF K is dissolved in 27 vs 28 L; accordingly, the rise in ICF [K] is less than 4%.

Because the ICF [K] rises, some K diffuses into the ECF to keep the [K] ratio constant (i.e., close to 4% greater than 4 mM or to less than 4.2 mM).

Since the vast bulk of K is in the ICF, you can ignore the quantity of K shifted in this calculation.

6. Since only 2% of K is in the ECF, at a constant RMP, 2% of the added K (70 mmol) remains in the ECF (i.e., 1.4 mmol), and this is distributed in an ECF volume of 14 L. Thus, the rise in [K] is only 0.1 mM.

ETIOLOGY OF HYPERKALEMIA

Excessive K Intake as a Cause for Hyperkalemia

CHRONIC SEVERE HYPERKALEMIA OCCURS WITH INCREASED K INTAKE ONLY IF K EXCRETION IS AT A REDUCED RATE.

The retention of 70 mmol of K produces a 0.1 mM rise in the plasma [K] in normals (see Answer 6 on page 257 for details). However, should a defect in K movement into cells be present, severe hyperkalemia could ensue. A very high K intake is rare, e.g., the use of large quantities of salt substitutes that contain 10–13 mmol K/g and the intake of K salts of organic acids (citrate) given to alkalinize the urine. Extremely high K intakes only lead to *chronic* hyperkalemia in patients with compromised K excretion. Finally, under special conditions, unusually large K loads can be administered IV (see page 259 for details).

K Shift From the ICF to the ECF as a Cause of Hyperkalemia

K SHIFTS FROM CELLS WITH HORMONE DEFICIENCIES, ACIDOSIS, OR CELL DAMAGE

Hormones

Insulin and β_2-adrenergic agents are the major hormones that cause K movement into cells. Hyperkalemia is seen if these hormones are relatively inactive.

Insulin. Insulin promotes K entry into cells by hyperpolarizing the cell membrane and by increasing the net anionic charge (organic phosphate accumulation). In acute experiments in which the basal insulin level was halved, the plasma [K] rose 0.5 mM. It is possible that even larger changes in the [K] could have occurred if insulin levels had fallen more.

Questions (Answers on page 259)

7. Under what circumstances do β-blockers cause *severe* hyperkalemia?
8. What are the major threats of hyperkalemia?

Unusually Large Intravenous K Loads. If KCl is added to an intravenous solution and this solution is not mixed adequately, it is possible to deliver a large K bolus to the heart with potentially disastrous effects. A large intravenous K load may also be given when large volumes of whole blood are transfused; the degree of this danger depends on the preservative, the temperature of the blood, the occurrence of cell lysis, and the duration of blood storage. Should the citrate that is used as an anticoagulant acutely lower the plasma ionized calcium concentration, the biologic response to hyperkalemia may be more severe.

In patients undergoing aortic surgery, rapid reperfusion after clamp removal can result in a sudden K bolus to the heart. An analogous situation may occur following renal transplantation because the transplanted kidney is perfused with a hyperkalemic solution during preservation. It should be appreciated that patients in these situations may have other problems that can contribute to the hyperkalemia.

Answers

7. There are several settings in which it is important to have the "K-shift defense" operating at its greatest efficiency (for example, following heavy physical exercise, trauma causing cell necrosis, a large K intake, renal failure, insulin plus aldosterone deficiencies, and after surgery). β-blockade at these times can cause severe hyperkalemia.

8. Severe hyperkalemia can cause arrhythmias or even cardiac arrest. This cardiotoxicity is more pronounced in patients with underlying cardiac dysfunction.

Hyperkalemia may be more harmful in patients with concomitant hypocalcemia (hypercalcemia counteracts the increased excitability in patients with hyperkalemia).

The K shift out of cells during the acidosis of diabetic ketoacidosis is not due to a pH change; this can be deduced from direct experiments and the time course of events (i.e., the plasma [K] falls 1–2 h after insulin was given; at this time there is little change in the acid-base status).

β_2-***Adrenergics.*** β_2-adrenergics lead to a K shift into cells; hence, β-blockers may cause hyperkalemia. However, these drugs raise the plasma [K] only a few tenths mM in normal subjects. (Note that β_1-blockers diminish renin release, and this can reduce aldosterone levels and thereby K excretion). In contrast, α-adrenergics have the opposite effect; they tend to increase the severity of hyperkalemia if there is already a stimulus to move K out of cells.

Aldosterone. Low levels of aldosterone do not cause serious hyperkalemia due to K shift out of cells; rather hyperkalemia is due to reduced renal K excretion.

Combined Hormone Deficiency. Should a patient have both hypoaldosteronism and low insulin levels, a much greater degree of hyperkalemia is seen. In a subject with a background of this bihormonal deficit, hyperglycemia can markedly exacerbate the degree of hyperkalemia; otherwise, hyperglycemia per se does not have a major effect on the plasma [K].

Questions (Answers on page 261)

9. What happens to the plasma [K] during exercise?
10. How low must the GFR be to cause hyperkalemia with a normal K diet (70 mmol/day)? (Assume normal aldosterone action.)

Acid-Base Factors

- Hyperkalemia is not caused by:
  - Respiratory acidosis
  - Ketoacidosis per se
  - Lactic acidosis
- If hyperkalemia accompanies chronic acidosis, look for a renal problem with K excretion.

Acute respiratory acidosis per se does not cause an appreciable rise in the plasma [K]. Similarly, metabolic acidosis due to the accumulation of lactic acid or ketoacids does not cause an appreciable K exchange with H^+ (insulin deficiency,

Answers

9. The plasma [K] may rise by several mM following exhausting exercise. Although all the mechanisms are not clearly defined, muscle cell depolarization yielding a high interstitial [K] that is washed into the circulation is the most likely cause. Cell damage (muscle, RBCs mechanically ruptured by persistent pressure on the soles of the feet) may contribute to this rise. In this situation, the hyperkalemia is particularly well tolerated by the otherwise normal subject, and its duration is very short-lived. β_2-adrenergic activity helps to minimize the degree and the duration of the hyperkalemia.

10. Normally, approximately 5–10% of the GFR reaches the cortical distal nephron. If aldosterone is acting, the [K] in the luminal fluid should be at least 7 times that in the plasma (assume 5 mM for this example); thus the luminal [K] should be 35 mM. Therefore, to excrete 70 mmol of K, 2 L of fluid must traverse this nephron segment. If 2 L must be delivered (representing 5% of the GFR), then a GFR of 40 L/day will permit the excretion of this daily K load (this is a conservative estimate, as we used only 5% of the GFR and a transtubular [K] gradient of only 7).

Three points should be obvious from this example: (a) Hyperkalemia does not become a problem until the GFR is reduced by at least 80% in an otherwise normal kidney; (b) low aldosterone bioactivity is required to produce significant hyperkalemia in patients whose GFR is greater than 20% of normal; (c) hyperkalemia occurs during low GFR states when the dietary K load is increased.

cell damage, or inadequate energy for the NaKATPase owing to hypoxia can cause a K shift to the ECF in these patients). In contrast, if a patient has metabolic acidosis due to $NaHCO_3$ loss, acute hyperkalemia can ensue. However, if the kidneys and adrenal glands are normal with respect to K handling, this hyperkalemia will not persist (see Chapter 9, Fig. 9–3, page 219).

Cell Damage

Since the ICF [K] is so high relative to that in the ECF, if an appreciable number of cells are damaged, hyperkalemia results. This is typically seen in patients suffering from trauma or those given cytotoxic treatment for neoplasms. The degree of hyperkalemia is worse if patients also have reduced K excretion or the hormonal deficiencies described above. Cell damage also occurs with hemolytic anemias, bleeding into the gastrointestinal tract, fulminant infections with white cell destruction, and rhabdomyolysis.

Although not technically an example of cell damage, hyperkalemia can be caused in special circumstances by a cellular K shift due to "destruction" of the resting membrane potential. Succinylcholine normally only depolarizes motor end-plates. However, if muscle is denervated and becomes "supersensitive" to depolarization, succinylcholine may depolarize the entire muscle, with the result being more severe hyperkalemia (instead of the 1 mM transient rise in the plasma [K]).

A list of drugs that can cause hyperkalemia classified according to mechanism of action is shown in Table 11–2, page 263.

Hyperkalemia Due to Reduced K Excretion in the Urine

Virtually all the K excreted entered the urine via the cortical distal nephron (see Chapter 9 for details). Two major renal factors are required: the actions of aldosterone (to reabsorb Na and raise the luminal fluid [K]) and an adequate volume delivered to this nephron site.

Laboratory Investigation of Reduced K Excretion. A reduced rate of K excretion is established by determining that K excretion is less than intake in a hyperkalemic subject. Usually, this means that the 24-hr K excretion rate is less

TABLE 11–2. Drugs That Can Cause Hyperkalemia

Drugs Containing K (only if renal function is compromised)

- e.g., KCl, other K salts

Drugs Causing a K Shift From ICF to ECF

- Hormone antagonists
  - β_2-Adrenergic blockers
  - α-Adrenergic agonists
  - Drugs that impair insulin release from B cells
- Cell depolarizers such as succinylcholine, digitalis overdose
- Drugs causing cell necrosis

Drugs That Interfere With K Excretion in the Urine

- Drugs causing renal failure
- Drugs interfering with aldosterone
  - Release from adrenal gland
    - β_1-Adrenergic blockers diminish renin release
    - Converting enzyme inhibitors (e.g., captopril)
  - Block aldosterone binding to its renal receptor
    - Spironolactone
    - Drugs causing interstitial nephritis
  - Postreceptor blockers
    - Na^+ channel blockers in the "cortical distal nephron" (amiloride)
    - Na^+ transport across the basolateral membrane

than 30 mmol. The causes of reduced K excretion include an extremely low GFR (marked ECF volume contraction with low distal delivery) and reduced aldosterone bioactivity (see Table 11–3A, page 265). Aldosterone bioactivity can be assessed by examining the transtubular [K] gradient in the cortical distal nephron in vivo (Fig. 9–4, page 233). The role of the adrenal gland can be assessed by hormone measurements (Tables 11–3A and 11–3B, page 265).

Simply examining the natural history of patients with aldosterone deficiency illustrates the problems associated with the clinical assessment of reduced K excretion. If the patient has chronic hyperkalemia and is in a steady state, the 24-hr urine K excretion equals that absorbed from the diet. Therefore, a 24-hr urine K excretion is not diagnostic unless it is interpreted in conjunction with the plasma [K]. There are no specific criteria to define how much K excretion should be expected for a given degree of hyperkalemia. However, using the urine [K] adjusted for medullary water abstraction and expressing this value relative to the plasma [K] offers the clinician an index of aldosterone action on the kidney.

$$\text{K excretion} = \underset{\text{(aldosterone)}}{[\text{K}]_{\text{urine}}} \times \underset{\text{(salt and water)}}{\text{Volume}_{\text{urine}}}$$

Plasma Aldosterone and Renin. In the hyperkalemic patient, failure to find a high aldosterone concentration suggests an adrenal or a renin problem. If the renin level is high, then the problem is adrenal, or there is a converting enzyme inhibitor present. In contrast, if the patient has a low or low-normal ECF volume and the renin level is not elevated, then the problem is renal in origin (Tables 11–3A and 11–3B, page 265). In some patients, there may be two renal problems: low renin release and low renal response to aldosterone.

SYMPTOMS AND SIGNS OF HYPERKALEMIA

Weakness. When hyperkalemia is present, the resting membrane potential is lower than normal. If this potential falls below the threshold potential, cells cannot repolarize. Patients complain of weakness, and ultimately paralysis

TABLE 11–3A. Factors to Evaluate in Hyperkalemic Patients With Low Aldosterone Bioactivity

Plasma Renin Concentration (while ECF volume is low)
Low
Problem with juxtaglomerular apparatus
High
Converting enzyme inhibitor
Adrenal gland problem
Low renal response to aldosterone

Plasma Aldosterone Concentration
Low
Low renin
Converting enzyme inhibitor
Adrenal gland problem
High
Renal problem (interstitial nephritis, Cl shunt, etc.)
K-sparing diuretics, aldosterone antagonists

Renal Response to Mineralocorticoids (physiologic dose)
Low Urine [K]
Renal problem
Very low distal Na delivery
K-sparing diuretics
High Urine [K]
Low aldosterone levels due to an adrenal problem

TABLE 11–3B. Causes of Hyporeninemia in Hyperkalemic Patients

Destruction of Juxtaglomerular Apparatus
Interstitial Nephritis
Infection
Drugs
- Nonsteroidal anti-inflammatory agents
- Certain antibiotics, including methicillin

Depositions
- Urate
- Amyloid

Diabetes Mellitus
Pharmacologic Blockade
β_1-Adrenergic blockers
Unknown Causes
Those resulting in Na retention and ECF volume expansion (e.g., chloride shunt disorder)

occurs. These symptoms are only evident with very severe hyperkalemia (almost 8 mM). This weakness is first evident in the lower extremities.

ECG Changes in Hyperkalemia. These changes are listed on page 267. The plasma [K] at which the ECG changes are seen should be taken as rough approximates only. These changes may be seen at a lower [K] if hypocalcemia, hyponatremia or acidosis is present; they are more dramatic if the changes in plasma [K] are acute.

TREATMENT OF HYPERKALEMIA

Urgency of treatment

1. ECG changes
2. Degree of hyperkalemia
3. Anticipated rise in plasma [K]

Since hyperkalemia can arise from many causes, there is no universal therapy for this electrolyte abnormality. We will assume for simplicity that the clinician has determined the cause for hyperkalemia, done all the diagnostic tests, and initiated specific therapy, where applicable. The urgency for treatment is dictated by two facts—the degree of elevation of the plasma [K] and the anticipated rate of K release from the ICF. The degree of abnormality in the ECG can help the clinician decide on the urgency of therapy because a major aim of treatment is to prevent cardiac arrhythmias. A list of the treatment modes is presented in Table 11–4.

Prevent a Further Rise in the Plasma [K]

K Intake. K intake should be as low as possible. Do not overlook the fact that certain medications are K salts (certain penicillin preparations, alkalinizing salts).

Prevent the Absorption of Dietary K. Certain resins bind K avidly enough to diminish the net absorption of K. In so doing, the cations, Na^+, Ca^{++}, or H^+ are displaced and absorbed. In general, this is not a problem. The major exchange resin used is Na polystyrene sulphonate; each g of resin may bind 1 mmol of K. Usually 15–30 g of resin are given q 6 h with 70% sorbitol to hasten transit to the colon because the colon is the major site of K binding (where K

TABLE 11–4. Therapy for Hyperkalemia

Therapy for Hyperkalemia
Stop K Intake
Increase K Loss From the ECF
Bind K in the GI Tract
K exchange resins
Oral resins have a slower onset of action
Resins given by enema have a faster onset of action
Shift K into Cells
Hormones
Insulin
Acid-base factors
Bicarbonate
Gain of ICF anions
Replace phosphate deficits
Promote Urinary K Loss
Ensure adequate distal Na and volume delivery to the cortical distal nephron (e.g., restore ECF volume deficits or give diuretics if appropriate)
Remove mineralocorticoid antagonists or K-sparing diuretics
Administer mineralocorticoids

ECG Changes During Hyperkalemia. Hyperkalemia produces characteristic ECG changes that roughly parallel the severity of the rise in plasma [K]. The earliest changes are an increase in T-wave amplitude, leading to tall, narrow, peaked, symmetrical T waves. With a greater degree of hyperkalemia, the R-wave amplitude decreases, the S wave increases and the S-T segment becomes depressed. The P-R, QRS, and Q-T intervals all become prolonged, and the P-wave duration increases, whereas its amplitude decreases. With extreme hyperkalemia, there is progressive QRS and T-wave widening. Terminally, ventricular tachyarrhythmias may be observed. All of the above changes are exaggerated by the presence of hypocalcemia and hyponatremia.

SUMMARY OF ECG CHANGES DURING HYPERKALEMIA

1. Increased T-wave amplitude.
2. Decreased R wave.
3. S-T segment depression.
4. Decreased P-wave amplitude.
5. Prolonged P-R, QRS, and Q-T.
6. Absent P waves.
7. Sine wave pattern of QRST.
8. Ventricular arrhythmias.

is secreted). Hence, oral resins are therapeutically effective only after many hours; in contrast, if the resins are administered by enema, they are effective much sooner. (Give 100 g of resin in as little water as needed to dissolve it [200 ml]; keep it in the colon for as long as possible.)

Questions (Answers on page 269)

11. Does aldosterone act only on the kidney and the ICF-ECF interface to lower the plasma [K]?

12. What is the lesion in this patient? An otherwise healthy patient presents with hyperkalemia associated with mild ECF volume *expansion* and hypertension, *metabolic acidosis* with a normal plasma anion gap, and a *low urine [K]*. No drugs were taken.

Promote K Loss in the Urine

Two factors are required for K loss in the urine: those that raise the [K] in the lumen of the cortical distal nephron (aldosterone) and those that increase flow through this nephron segment. If the patient has a low aldosterone level, administer a physiologic dose (0.05 mg) of 9α-fluorohydrocortisone. If the urine [K] (adjusted for medullary water reabsorption) does not rise in 4 hr to at least 6 × plasma [K], you can quadruple the dose and follow the same parameters. Aldosterone action requires a lag period of up to 2 hours, and the maximum biological effect may take days to develop. If the patient is taking drugs that interfere with aldosterone action (K-sparing diuretics, competitive inhibitors), these must be discontinued. Reduced K excretion can occur if the volume of fluid reaching the cortical distal nephron is small. Suspect this in patients with ECF volume contraction or a low cardiac output due to heart problems. In the former case, saline administration is appropriate, whereas digitalis and/or a loop diuretic may help patients in the second category.

Prevent K Loss from the ICF

This is an important form of therapy in certain cases. If a patient is in a catabolic state, this should be reversed. Among the drugs to consider are those that cause cell lysis (antican-

Answers

11. No. Aldosterone also acts on the colon to promote fecal K loss, but this is only quantitatively important in patients with chronic renal failure in whom almost half of the daily K load is excreted in this manner.

The mechanisms involved in gastrointestinal K loss are analogous to those in the cortical distal nephron. However, glucocorticoids also promote K loss via the colon.

12. Hyperkalemia should lead to aldosterone release and enhanced K excretion. This did not occur. Therefore, either aldosterone is absent or not working. The absence of aldosterone as a primary lesion leads to Na loss and ECF volume *contraction*. Therefore, this patient does not have typical aldosterone deficiency because the ECF volume is expanded. If aldosterone is present and preventing Na loss, why does it not lead to K loss? Recall that a major action of aldosterone is to promote cortical distal nephron Na^+ reabsorption (opens the Na^+ channel). In this case, if the accompanying anion Cl^- were reabsorbed in parallel, there would not be a depolarization of the luminal membrane, resulting in K^+ and H^+ secretion; the latter (together with a hyperkalemia-induced inhibition of ammonium production) caused the metabolic acidosis. The lesion is the "chloride-shunt disorder."

cer medications, drugs causing hemolysis, etc.) and those that promote catabolism (antimetabolites, etc.). Another potential for cell lysis that should be considered is bleeding into the gastrointestinal tract.

However, since 1 L of blood has only 0.4 L of RBC (ICF K = 100 mM), the quantity of K released is only 40–45 mmol/L of blood digested.

Although it may be difficult to stop all of the above, their presence should dictate a more active role in promoting K loss on a longer-term basis. Without a rapid reversal in etiology, the patient should be considered for dialysis before life-threatening hyperkalemia ensues.

Shift K Into Cells

There are two major options here: raise the ECF [HCO_3] or induce a K shift into cells by hormonal action (see page 271 for time course).

$NaHCO_3$ Therapy. If the patient had metabolic acidosis due to $NaHCO_3$ loss, the acidemia could have led to a K shift into the ECF. This can be reduced by administering sufficient $NaHCO_3$ to bring the blood pH towards normal. Even in patients with a normal blood pH, HCO_3 therapy (50 mmol over 5 min, repeat 30 min later if necessary) can be effective in lowering the plasma [K]. The onset of K lowering takes at least 30 min. The dangers of this therapy relate to the Na load given (ECF volume expansion) and to potential problems of alkalemia (tetany, etc.). Finally, this effect is limited in scope and must be considered a temporary measure (lasts up to a few hr); nevertheless, the effects can be dramatic.

Hormonal Therapy. Two major hormones need to be considered—insulin and β_2-adrenergics. If the patient is severely hyperglycemic, insulin can have a dramatic effect in lowering the plasma [K]. Insulin therapy can also be dramatic in diabetics with aldosterone deficiency.

To lower the plasma [K], large doses of insulin may be necessary, but do not rely on this as a long-term therapy. As an example, giving 500 ml of 10% glucose in water stimulates endogenous insulin release; if diabetes mellitus is present, give 1 unit of insulin for every 3 g of glucose administered.

Time Course for Effectiveness of Therapeutic Agents

1. **Seconds to minutes**
   Antagonist to cardiac effects of hyperkalemia such as intravenous calcium gluconate

2. **30 min–1 hr**
   $NaHCO_3$
   ECF volume expansion

3. **1–4 hr**
   Insulin
   Aldosterone agonists
   Rectally administered K-binding resins

4. **Greater than 6 hr**
   Orally administered K-binding resins

5. **Dialysis**
   Apart from the dialysis membrane and the flow rate past this membrane, the major factors influencing the dialysis of K is the [K] gradient: the higher the plasma [K] and the lower the bath [K], the greater the rate of K removal. If the plasma [K] falls (e.g., if glucose is added to the dialysis fluid), K removal by dialysis might then be less.

β_2-adrenergics shift K into cells. Therefore, β-blockers should be discontinued. There is not much therapeutic leverage in administering β_2-adrenergics to lower the plasma [K].

While not directly related to hormonal or acid-base effects, should there be a lack of an essential ICF constituent such as phosphate, magnesium, etc., these should be given to promote anabolism and a subsequent K shift into cells.

Antagonize the Effects of K

The administration of calcium salts can decrease membrane excitability and thereby protect against the results of hyperkalemia. This effect begins within minutes but is relatively short-lived; it "buys 60 minutes" before other forms of therapy (insulin, $NaHCO_3$) can lower the plasma [K]. The usual dose is 10 ml of 10% calcium gluconate infused over 2–3 min. This dose can be repeated in 5–10 min if ECG changes persist (note the danger of hypercalcemia-induced digitalis toxicity if cardiac glycosides are being given).

Remove K by Dialysis

When renal function cannot eliminate sufficient K, dialysis therapy should be considered. There are two major indications for K removal by dialysis: either the plasma [K] is dangerously high (or there are ECG changes and a somewhat lower plasma [K]) or an ongoing K shift from cells is anticipated. Hemodialysis is preferred over peritoneal dialysis, as it is more efficient at removing K. Although there will be some variation due to membranes and flow rates, expect to remove close to 30–50 mmol of K/hr by hemodialysis and 15–25 mmol of K/hr by peritoneal dialysis.

Suggested Reading

Adrogue HJ and Madias NE: Changes in plasma potassium concentration during acute acid-base disturbances. Am J Med 71:456–467, 1981.
Adrogue HJ, Wilsin H, Boyd AE III, et al: Plasma cell-base patterns in diabetic ketoacidosis. N Engl J Med 307:1603–1610, 1982.
DeFronzo RA: Hyperkalemia and hyporeninemic hypoaldosteronism. Kidney Int 17:118–134, 1982.
DeFronzo RA, Shewrwin RS, Dillingham M, et al: Influence of basal insulin and glucagon secretion on potassium and sodium metabolism. J Clin Invest 61:472–482, 1978.
Gordon RD: Syndrome of hypertension and hyperkalemia with normal glomerular filtration rate. Hypertension 8:93–102, 1986.
Nerup J: Addison's disease—clinical studies. A report of 108 cases. Acta Endocrinol 76:127–141, 1974.
Pedro Ponce S, Jennings AE, Madias NE, et al: Drug-induced hyperkalemia. Medicine 64:357–370, 1985.
Szylman P, Better OS, Chaomowitz C, et al: Role of hyperkalemia in the metabolic acidosis of isolated hypoaldosteronism. N Engl J Med 294:361–365, 1976.
Williams GH: Hyporeninemic hypoaldosteronism. N Engl J Med 314:1041–1042, 1986.

12

METABOLIC AND PHYSIOLOGIC ASPECTS OF HYPERGLYCEMIA AND KETOACIDOSIS

THE BIG PICTURE: Hyperglycemia

Relative insulin deficiency is required to produce chronic severe hyperglycemia.

Severe hyperglycemia is almost always accompanied by a low GFR. The degree of hyperglycemia is also influenced by glucose intake.

The ECF volume is low in severe hyperglycemia, unless chronic renal failure is present.

Hyperglycemia causes muscles to shrink, liver to swell, and brain cell size to remain close to normal. The plasma [Na] falls 1.35 mM for every 100 mg/dl (5.5 mM) rise in blood [glucose].

Treatment: **Insulin can be withheld acutely in the absence of acidosis or hyperkalemia. Saline is needed to reexpand the ECF volume and to promote glucosuria. Give K for initial hypokalemia or 1 hr after insulin acts; hypoglycemia may occur 6 hr after insulin acts. Treat the underlying disorders.**

Ketoacidosis is a common complication of IDDM and is much less common in non-IDDM.

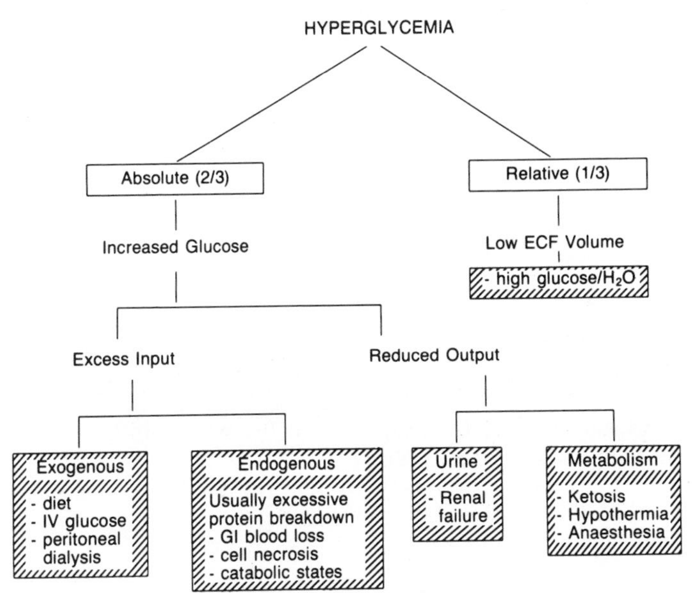

Flow Chart 12–A. Approach to hyperglycemia. The information required to determine the cause of hyperglycemia appears in the open boxes, and the final diagnoses are shown in the shaded boxes. To raise the plasma glucose concentration, there must be either more glucose input into the glucose space (an "absolute" rise in glucose, which accounts for ⅔ of the rise in glucose concentration in most cases) or a diminished volume in which to distribute that glucose (a "relative" rise in glucose concentration). Usually, both mechanisms contribute to hyperglycemia. In the reduced glucose output category, low renal excretion of glucose, owing to a low GFR, is most important. For glucose metabolism, fat oxidation or low rates of O_2 consumption would diminish the glucose oxidation rate.

ILLUSTRATIVE CASE

An elderly woman received a diuretic for hypertension. Urine output had increased markedly in the past few days and she felt poorly. She diminished the intake of food and stopped all medications. She was brought to the hospital because she was confused. Physical examination revealed marked ECF volume contraction and confusion. Plasma values are summarized below:

Na	126 mM	Glucose	1000 mg/dl (55 mM)
K	4 mM	BUN	42 mg/dl (15 mM)
Cl	80 mM	pH	7.40 ([H^+] 40 nM)
HCO_3	24 mM	Pa_{CO_2}	40 mm Hg

Serum ketones: Weakly positive

Questions (Answers on page 277)

1. Why is the plasma [Na] low?
2. Is the ICF volume high, low, or normal?
3. In quantitative terms, to what degree should the plasma [Na] fall, owing only to the water shifts associated with hyperglycemia?
4. Is the total body K content normal?
5. Is there an acid-base disorder?

APPROACH TO SEVERE HYPERGLYCEMIA (see Flow Chart 12–A, page 275)

CHRONIC SEVERE HYPERGLYCEMIA IS ALWAYS ASSOCIATED WITH RELATIVE INSULIN LACK.

Virtually all patients with severe hyperglycemia (>500 mg/dl, 28 mM) have relative insulin deficiency. The rare exception to this rule is after the sudden infusion of glucose by intravenous or peritoneal routes. A simple approach to a patient with severe hyperglycemia is to try to determine the relative importance of increased glucose input and reduced glucose output as causes for the hyperglycemia. The major source for a large increase in glucose input is an exogenous one, whereas glucosuria and "insulin-insensitive" glucose metabolism are the two major paths for glucose output. Both of these are examined below.

Answers

1. Plasma [Na]. This is the Na:H_2O ratio. It is low because hyperglycemia causes water exit from muscle ICF to ECF (9 × 1.35 mM = 12 mM). Therefore, hyperglycemia can account for the fall in the plasma [Na] from 138 to 126 mM. In addition, the Na content declined because of Na excretion in the urine due to the diuretic and the osmotic diuresis. A fall in plasma [Na] occurs if there is pseudohyponatremia (hyperlipemia). Polydipsia may also cause hyponatremia in conjunction with ADH release due to ECF volume contraction.

2. The ICF Volume. The ICF volume of some organs is increased, in others it is normal, whereas in yet others, it is reduced.

Muscle (20 L ICF). Glucose is restricted to the ECF in osmotic terms, at least, because the ICF [glucose] is in the uM range. Glucose thus acts like Na to draw water out of muscle cells, and the muscle ICF volume declines during hyperglycemia (the ECF is hypertonic when one considers the [Na] plus [glucose].

Liver (0.8 L ICF). Glucose crosses the liver cell membrane, so that the [glucose] is equal in the hepatic ICF and ECF; hence, it makes no direct contribution to water shifts there. However, hyponatremia shifts water into the hepatic cells and they swell.

Brain Cells (1 L ICF). The higher [glucose] in the ECF draws water out of these cells. Water enters, owing to hyponatremia. In addition, CNS cells tend to regulate their volume independently. Therefore, brain cell volume may be close to normal before therapy.

3. Quantitation. The plasma [Na] falls by 1.35 mM/100 mg% rise in [glucose] (or 1 mM fall in [Na]/3.7 mM rise in glucose).

4. Total Body K Content. Since only 2% of K is normally in the ECF, the plasma [K] is a poor indicator of the total body K content. Insulin deficiency causes K to shift from the ICF to the ECF. The K shift is larger if tissue catabolism and/or necrosis is present. However, much K is lost in the urine, owing to the osmotic diuresis (aldosterone levels are high due to the ECF volume contraction). Thus, these patients have a low total body K content, but this K deficit is not reflected by the degree of hypokalemia.

5. Acid-Base Disorder. Despite the fact that the pH, Pa_{CO_2} and [HCO_3^-] are all normal, there are acid-base disorders present. We can deduce that a form of metabolic acidosis is present from the wide plasma anion gap of 22 (most likely β-hydroxybutyric acidosis). In addition, there must also be a process that raised the plasma [HCO_3^-] (metabolic alkalosis due to the diuretic), as the plasma [HCO_3^-] is normal despite the wide anion gap.

WATER SHIFTS DUE TO HYPERGLYCEMIA

Since glucose remains largely in the ECF, hyperglycemia induces a water shift from the ICF (Fig. 12–1, page 279). In summary, muscle ICF volume decreases, liver ICF volume expands, and the brain size remains close to normal. The plasma [Na] declines 1.35 mM for every 100 mg/dl (5.5 mM) rise in the plasma glucose concentration.

GLUCOSE POOL SIZE

With severe hyperglycemia, the total glucose pool is the blood glucose concentration × the volume of distribution of glucose (the ECF volume [14 L] plus an additional 3–4 L of ICF). Since severely hyperglycemic patients usually have marked ECF volume contraction (owing to the osmotic diuresis), the total glucose content of a 70 kg person with a blood glucose of 1000 mg/dl (10 g/L) is 150 g (833 mmol) (15 L × 10 g/L).

GLUCOSE INPUT

Glucose can be derived from endogenous or exogenous sources. The endogenous sources are glycogen, protein, and glyceride-glycerol (see page 283 for more detail).

Exogenous Glucose. A typical diet supplies close to 225 g of glucose (or its precursor). However, most severely hyperglycemic patients are quite ill and do not consume this quantity of glucose. We shall assume a glucose input of 225 g for subsequent calculations but recognize that this can range from zero to a very high intake.

Question (Answer on page 279)

6. What weight of muscle must be broken down to yield 75 g of glucose? This question illustrates the marked catabolism associated with hyperglycemia.

GLUCOSE OUTPUT

There are two major sites of glucose loss: glucose metabolism not requiring insulin (CNS) and glucosuria.

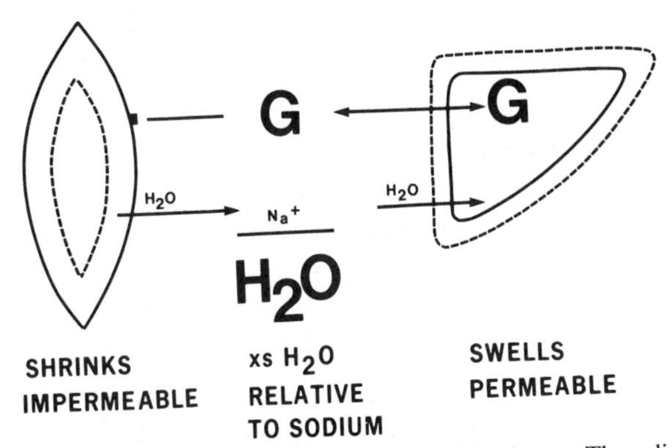

Figure 12–1. Influence of hyperglycemia on ICF volumes. The solid lines represent the original volume, and the dashed lines the volume after water shifts. The "G" in bold type represents hyperglycemia. The term "impermeable," under the muscle on the left, and the term "permeable," under the liver to the right, describe glucose movement in each organ.

The concentration of glucose in muscle cells is always so low that hyperglycemia causes water to move from its ICF to the ECF down the osmotic gradient. In contrast, the ICF glucose concentration in liver is equal to that in the ECF. Therefore, water movement across liver cell membranes responds to changes in the [Na]. xs = excess.

Answer

6. Since only 60% of the weight of protein can be converted to glucose, then 125 g of protein (75 g glucose/0.60) must have been catabolized. However, roughly 80% of the weight of lean body mass is water. If all the remainder was protein, then 625 g (1.3 lb) of muscle was catabolized. To place this in proper perspective, it would not be unreasonable to find 150 g of glucose lost in the urine each day and a body glucose pool size that is 75 g above normal. Therefore, you can identify the catabolism of at least 3 lb of lean body mass (if dietary glucose were nil). Hence, this is a very catabolic situation.

When therapy is considered, all the ICF constituents lost during the catabolism (K, Mg, phosphate, protein, etc.) must ultimately be replaced.

Glucose Metabolism. In the absence of insulin, the CNS oxidizes 125 g of glucose to CO_2 and the rest of the body perhaps another 25 g. Thus, glucose oxidation to CO_2 is only 6 g/hr.

Glucosuria. This can be the major pathway for glucose removal. If you consider that a normal kidney reabsorbs 350 g of glucose per day, then the filtered load of glucose that exceeds this amount is excreted in the urine (Table 12–1, page 281).

GLUCOSE BALANCE IN SEVERE HYPERGLYCEMIA

> HYPERGLYCEMIA IS MAINLY A FUNCTION OF THE GFR; IT IS ALSO INFLUENCED BY GLUCOSE INTAKE.

If glucose input was 325 g (100 g/day endogenous and 225 g exogenous) and 150 g of that glucose were obligatorily oxidized to CO_2 per day, then 175 g of glucose must be excreted to achieve glucose balance. Since the kidney reabsorbs 350 g of glucose per day, then glucose balance is achieved by filtering 175 g + 350 g (525 g) of glucose (Table 12–1, page 281). When the blood [glucose] is 1000 mg/dl (10 g/L), the GFR must be close to 50 L/day (25% of normal) to filter 525 g of glucose. Hence, severe steady-state hyperglycemia requires that the GFR be significantly reduced.

Questions (Answers on page 281)

7. How much glucose and electrolytes are contained in 1 L of osmotic diuresis?

8. A patient presents with a blood [glucose] of 1000 mg/dl. The ECF volume was *expanded*. No exogenous glucose was ingested, and ketosis is absent. What lesion is present in this patient that can explain the above findings?

9. What problems could lead to a higher blood [glucose] in a patient with insulin deficiency?

10. Does a patient with a blood [glucose] of 1000 mg/dl and a plasma [K] of 5 mM have a K deficit? If so, how big a deficit is present?

TABLE 12–1. Glucose Excretion at Various Blood [Glucose] and GFR*

Blood [Glucose] (mg/dl)	Glucose Excretion (g/day)		
	GFR = 180 L/day	*GFR = 90 L/day*	*GFR = 45 L/day*
200	10	0	0
500	550	100	0
1000	1450	550	100
1500	2350	1000	325

*GFR = glomerular filtration rate. Calculations are based on a rate of renal glucose reabsorption of 350 g/day. Tubular damage decreases glucose reabsorption and thereby increases glucosuria. At a net glucose production rate of 100 g/day, the GFR would have to be 50% of normal if the blood [glucose] were 500 mg/dl or 25% of normal if the blood [glucose] were 1000 mg/dl.

Answers

7. Each 50 g (278 mmol) of urinary glucose causes the excretion of 1 L of urine and 50 mmol of Na plus 20–30 mmol of K.

8. Severe hyperglycemia requires either excessive glucose input (not present) or diminished glucose loss. Since ketosis is absent, glucose oxidation to CO_2 is not diminished; therefore, this diabetic patient must have reduced glucose loss by the other major pathway—the urine; therefore, the patient has renal failure. Hyperglycemia causes water to shift from the ICF of muscle to the ECF; this contributes to the ECF volume expansion in this case.

9. A higher blood [glucose] is present if glucose input increases or if glucose output decreases. A higher input could come from the diet (or excessive protein breakdown). A reduced glucose output could occur if CNS glucose oxidation were diminished (ketoacidosis) or if glucosuria declined (renal failure).

10. Since severely hyperglycemic patients are very catabolic, there are 2 processes elevating the ECF [K] at the expense of ICF K: severe catabolism and insulin deficiency. Hence, the plasma [K] is a poor indicator of the degree of K deficit. As a rough index, expect a K deficit of 5–10 mmol/kg body weight in diabetic ketoacidosis. However, you cannot replace this entire K deficit without supplying the other ICF constituents (mainly phosphate). At least half of the K administered will appear in the urine.

TREATMENT OF THE SEVERELY HYPERGLYCEMIC PATIENT

- REEXPAND THE ECF VOLUME WITH SALINE
- WATCH OUT FOR HYPOKALEMIA AT 1–2 HR
- THE BLOOD GLUCOSE CONCENTRATION SHOULD FALL AT 100 mg/dl/hr (5.5 mM/hr)
- REPLACE PHOSPHATE DEFICITS

The aims of therapy are to lower the blood glucose concentration at a reasonable rate, while trying to avoid serious hypokalemia. After 6–8 hours, there may be a threat of hypoglycemia.

It is important to suspect, investigate, and treat any underlying disorder.

Saline Administration. Unless hyponatremia is marked or the patient is in shock, infuse half-normal saline rapidly (0.5–1 L/hr) initially. Monitor both the ECF volume (slow the infusion when the ECF volume approaches normal) and the plasma [Na] (to determine the infusate [Na]). Anticipate an increase in [Na] as the [glucose] falls and water enters the ICF.

Question (Answer on page 283)

11. By what mechanisms does saline administration cause the blood glucose concentration to fall?

Insulin. Insulin must be given if there is ketoacidosis or severe hyperkalemia; a drawback to its administration is the risk of severe hypokalemia. There is debate as to whether insulin is absolutely required to lower the blood glucose concentration early during therapy. If there is no acid-base or K threat, insulin may be withheld initially to prevent severe hypokalemia. If administered, watch out for delayed (8+ hr) hypoglycemia, and give glucose once the blood glucose concentration approaches 270 mg/dl (15 mM).

K. K should be given to all hypokalemic patients at the outset. If insulin is given, anticipate a fall in plasma [K] after 90 min (when insulin acts). Each L of urine should contain 20–30 mmol of K during the osmotic diuresis.

Since these patients are also phosphate deficient, it would be advantageous if some of the K administered were the phosphate salt (no faster than 6 mmol/hr).

ENDOGENOUS GLUCOSE PRODUCTION (100 g/day)

Glycogen. Muscle glycogen (200 g) is not readily catabolized other than to support anaerobic glycolysis. In contrast, liver glycogen (100 g in the fed state) can be released as glucose. In chronic insulin-deficit states, this glycogen has already been used; additional new glucose formation is unlikely from this source.

Protein. About 75 g of glucose is formed each day.

Glyceride-Glycerol. $<$25 g of glucose is formed per day.

Answer

11. Saline administration increases the ECF volume toward normal. This can cause the blood [glucose] to fall for 3 reasons:

a. *Dilution* of the glucose pool (usually this causes a 25–33% fall in the [glucose] without a change in *glucose content*.

b. *Glucosuria* increases because reexpansion of the ECF in these patients raises the GFR on a prerenal basis. You can see from Table 12–1, page 281, that this can cause a major fall in the glucose pool size (this action does not require insulin).

c. *Insulin* release from the B cells of the pancreas may be inhibited by α-adrenergic actions—the adrenalin being released in response to ECF volume contraction.

Suggested Reading

Arieff AI and Carroll HJ: Nonketotic hyperosmolar coma with hyperglycemia: clinical features, pathophysiology, renal function, acid-base balance, plasma cerebrospinal fluid equilibria and the effects of therapy in 37 cases. Medicine 51:73–94, 1972.
Bjorntorp P and Sjostrom L: Carbohdyrate storage in man: speculations and some quantitative considerations. Metabolism 27:1853–1865, 1978.
Halperin ML, Goldstein MB, Richardson R, et al: Quantitative aspects of hyperglycemia in the diabetic: a theoretical approach. Clin Invest Med 4:127–130, 1980.
Kurtzman NA, White MG, Rogers PW, et al: Relationship of sodium reabsorption and glomerular filtration rate to renal glucose reabsorption. J Clin Invest 51:127–133, 1972.
West ML, Marsden PA, Singer GG, et al: Quantitative analysis of glucose loss during acute therapy for hyperglycemic hyperosmolar syndrome. Diabetes Care 9:465–471, 1986.

THE BIG PICTURE: Alcohol Metabolism

The acid-base disorders caused by ethanol abuse are largely the result of vomiting and ethanol metabolism. The former produces Na, K, and Cl loss and metabolic alkalosis. Ethanol metabolism in the liver leads to ketoacidosis, lactic acidosis, and inhibition of glucose production.

Insulin release is inhibited by the α-adrenergic response to ECF volume contraction.

Therapy consists of K replacement and ECF volume reexpansion (NaCl). Avoid problems relative to B-vitamin deficiencies and phosphate depletion. Give glucose only to avoid hypoglycemia.

Do not overlook complications or underlying disorders.

ILLUSTRATIVE CASE

A 26 year old man had consumed an excessive quantity of alcohol in the past week; in the last 2 days he had eaten little and vomited on many occasions. He had no history of diabetes mellitus. Physical examination revealed marked ECF volume contraction. Alcohol was detected on his breath.

Blood		**Plasma**	
Glucose	90 mg/dl (5 mM)	Na	140 mM
BUN	28 mg/dl (10 mM)	K	3.0 mM
pH	7.30 ($[H^+]$ 50 nM)	Cl	93 mM
Pa_{CO_2}	30 mm Hg	HCO_3	15 mM
Ketones:	Strongly positive		

Questions (Answers, page 285)

12. What is the total body Na content?
13. Is there relative insulin deficiency?
14. Why is the patient hypokalemic? To what degree is the K depleted?
15. What acid-base diagnoses can you detect?
16. What are your primary considerations for therapy?
17. The plasma osmolality was 350 mOsm/kg H_2O. Is the ICF volume high, low, or normal?

Answers

12. Since the ECF volume is markedly contracted, there is a large Na deficit.

13. Yes, ketoacidosis signals relative insulin deficiency. Hyperglycemia is not present because of the inhibition of gluconeogenesis in the liver by ethanol.

14. The patient is severely K-depleted because hypokalemia is present in the face of relative insulin deficiency. Vomiting leads to large renal K loss (see page 237).

15. The acidemia, low [HCO_3^-] and wide plasma anion gap all indicate that metabolic acidosis is present. This is most likely due to ketoacidosis plus some lactic acidosis. However, metabolic alkalosis due to vomiting is also present, as the rise in anion gap (32 − 12 or 20 mEq/L) is much greater than the fall in the plasma [HCO_3^-] from normal (25 − 15 = 10 mM).

No respiratory acid-base disorder is present.

16. a. Isotonic saline to reexpand the ECF volume.
    b. KCl (40 mM) to replace the K deficit.
    c. B vitamins to replace nutritional deficits.
    d. Phosphate should be given if the patient is severely hypophosphatemic.

Do not give insulin, as the patient may suffer from acute hypokalemia (insulin is released from B cells once the ECF volume is reexpanded).

There is controversy concerning glucose administration; it should only be given to prevent hypoglycemia, in our opinion.

17. The plasma osmolality was 350 mOsm/kg H_2O, whereas the calculated value was (2[Na] + mM urea + mM glucose) 310 mOsm/kg. The difference is mostly due to the presence of 40 mM ethanol. Since the ethanol concentration (like that of urea) is equal in the ICF and ECF, it does not produce a water shift. This osmolal gap is helpful to detect ethanol as well as other low molecular weight uncharged molecules such as methanol, ethylene glycol, etc. (see also page 76).

The normal [Na] (and [glucose]) suggests that the ICF volume is normal.

18. Should alcohol not be present, the patient will still have ketoacidosis owing to lipid mobilization, but lactate levels will not be elevated, and the [glucose] will be higher.

APPROACH TO THE PATIENT WITH ACID-BASE DISORDERS RELATED TO ALCOHOL

The factors to evaluate on presentation are an estimate of recent alcohol intake, the reason for the marked ECF volume contraction (usually vomiting), and the nature of the metabolic defects. Suspect a deficit in Na, K, and Cl; these patients are also phosphate-depleted and may also suffer from vitamin deficiences. The major threats to life are shock (ECF volume contraction), hypokalemia (arrhythmias), and complications related to the underlying disorder.

Although acid-base disturbances are present, these are not life-threatening; neither do the variations in plasma [glucose] pose a major threat, unless very low values are observed and ketosis is absent.

TREATMENT

Treatment should, therefore, be to restore K deficits early if hypokalemia is present, to restore ECF volume aggressively, and to give some phosphate early in the course of therapy. B vitamins should also be added to the initial IV solutions.

Glucose should only be given to prevent hypoglycemia, and insulin need not be given unless acidemia is significant.

Watch out for complications (e.g., aspiration pneumonitis) and underlying disorders (e.g., acute pancreatitis). The electrolyte and acid-base abnormalities are largely corrected within 24 hr, but long-term therapy is required to correct completely all electrolyte and nutritional deficiencies.

Question (Answer on page 285)

18. If ethanol intake was discontinued 24 hr previously, what changes might you observe in the metabolic picture?

Suggested Reading

Halperin ML, Hammeke M, Josse RG, et al: Metabolic acidosis in the alcoholic: a pathophysiologic approach. Metabolism 32:308–315, 1983.

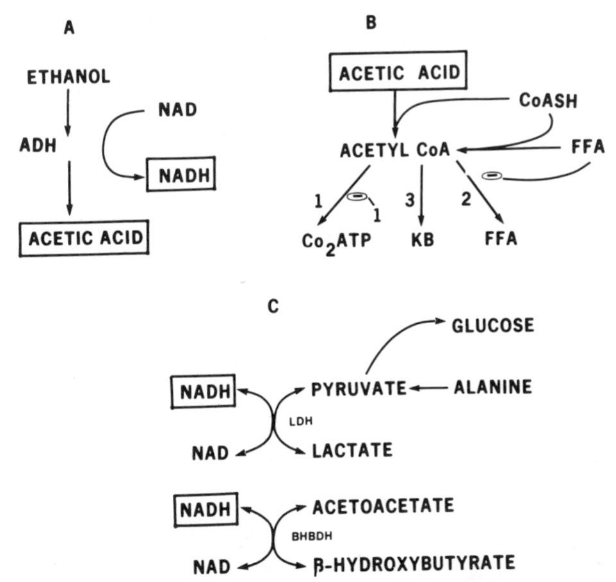

Figure 12–2. The biochemistry of ethanol metabolism. *A*, Alcohol dehydrogenase (ADH) initiates ethanol metabolism in the cytosol of the liver. It generates NADH and acetaldehyde. Acetaldehyde dehydrogenase (*lower arrow*) converts acetaldehyde + NAD to acetic acid + NADH in hepatocyte mitochondria. *B*, Acetic acid is converted to acetyl CoA, which has 3 fates: (1) oxidation to CO_2 and ATP if ATP is needed; (2) conversion to free fatty acids (FFA) only in the fed state; (3) conversion to ketone bodies (KB) in the fasted state. *C*, A high NADH diverts pyruvate to lactate (via LDH) and causes hypoglycemia when gluconeogenesis is absolutely required (chronic starvation). In the absence of ethanol, both lactate accumulation and low glucose production rates are corrected; however, ketoacids are produced (via BHBDH) from free fatty acids (FFA) as long as there is insulin lack.

Pathophysiology of Alcohol Abuse (Fig. 12–2)

The consumption of a large quantity of ethanol may provoke gastric irritation and vomiting. The secondary events are Na loss (ECF volume contraction), which leads to aldosterone release (renal K loss), and the development of metabolic alkalosis. The release of α-adrenergics as a consequence of ECF volume contraction inhibits insulin release from B cells, promoting fat mobilization and thereby ketoacidosis. Specific metabolism of ethanol also leads to ketoacid production; in addition, the NADH so formed diverts carbon destined for glucose into lactic acid, yielding acidosis and hypoglycemia.

13

APPLICATION OF PHYSIOLOGIC PRINCIPLES TO CLINICAL DIAGNOSIS

DIFFERENTIAL DIAGNOSIS OF POLYURIA

The approach to polyuria is shown in Flow Chart 13–A, page 289, and illustrated in the following case example:

A 47 year old female consumed a considerable quantity of methanol. As soon as she was brought to the hospital, ethanol was administered and hemodialysis instituted. Recovery was complete, but 24 hr after hemodialysis, polyuria was noted (0.7 L/hr) and this persisted for the next 5 hr. Physical examination on admission revealed only mild periorbital edema and hypoactive bowel sounds. At no time was the ECF volume contracted. Over the period of observation, she received 75 mM saline intravenously at a rate of 10 ml/min. Laboratory values are summarized below:

	On Admission	After Dialysis	During Polyuria
Plasma			
Na (mM)	141	139	141
K (mM)	3.1	4.0	4.0
Cl (mM)	96	94	98
HCO_3 (mM)	29	31	32
pH	7.45	7.49	7.49
P_{CO_2} (mm Hg)	40	41	41
Osm (mOsm/kg H_2O)	391*	282	290
Urine			
Na (mM)	—	198	92
K (mM)	—	30	16
Cl (mM)	—	154	48
Osm (mOsm/kgH_2O)	—	441	290
pH	—	—	8

*Methanol = 93 mg/dl (29 mM), ethanol = 327 mg/dl (71 mM), glucose = 90 mg/dl (5 mM), urea = 9 mg/dl (2.7 mM), creatinine = 0.8 mg/dl (69 uM).

Question (Answer on page 289)

1. What is the cause of the polyuria in this case?

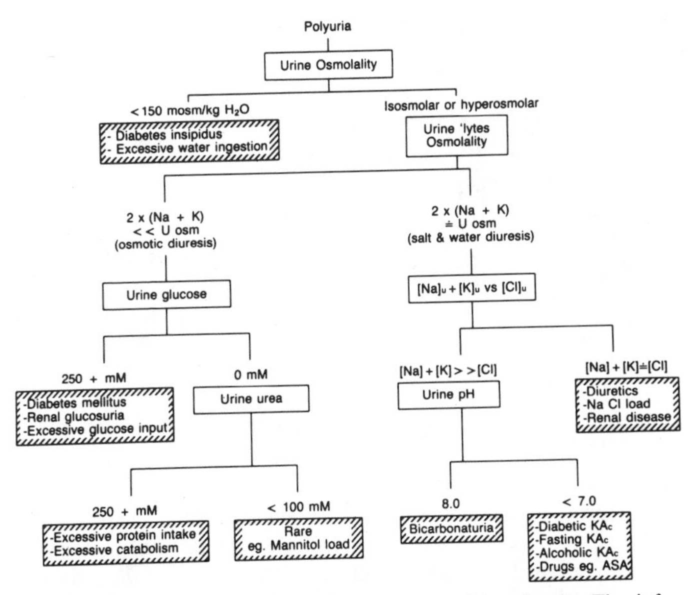

Flow Chart 13–A. The approach to patients with polyuria. The information required to determine the cause of polyuria is shown in the clear boxes, and the final diagnoses are shown in the shaded boxes.

Answer

1. Since the urine osmolality during the polyuria was 290 mOsm/kg H_2O, a water diuresis was unlikely. The next step was to discover which osmole was present in the urine. Since 2 × the sum of the [Na] + [K] in the urine accounted for most of the urine osmolality, glucose- or urea-induced osmotic diuresis was ruled out.

Diuretics, excessive salt and water intake, and intrinsic renal disease leading to diminished reabsorption of filtrate, are the most likely causes for a salt and water diuresis. The simplest step to resolve this differential diagnosis is to compare the urine [Cl] to the [Na] + [K]. In this case, the [Cl] was much lower than the [Na] + [K], indicating the presence of another anion; since the urine pH was close to 8, that anion was HCO_3. Furthermore, since the plasma [HCO_3] did not fall during the polyuria, there was HCO_3 input that was nonrenal in origin. The source of HCO_3 was unlikely to be vomiting because the ECF volume was not contracted. Since no $NaHCO_3$ was administered, suspect the presence of an ileus (containing $NaHCO_3$-rich fluid) early in the course (remember the hypoactive bowel sounds), which was now resolving, leading to $NaHCO_3$ reabsorption. With this diagnosis, no specific therapy was required.

CLUES FROM URINE ELECTROLYTES (Table 13–1, page 291)

> **URINE ELECTROLYTES**
>
> No normal values—evaluate excretion relative to need at that time.
>
> - Na and Cl: Assess relative to ECF volume and the possible presence of diuretics
> - K: Evaluate relative to plasma [K] and ECF volume. Look at urine [K] and excretion independently
> - HCO_3^- and NH_4^+: Examine relative to plasma acid-base status
> - Osmolality: Examine relative to plasma [Na]

Normal Values. Since normal subjects are in electrolyte and water balance, they excrete what they ingest minus nonrenal losses. Therefore, there is no meaningful normal range; *you must assess the urine electrolytes relative to the clinical setting*.

Na. The urine should be Na-free when the ECF volume is contracted. The only exceptions are if another substance obligates the loss of Na. The first compounds to think of are anions that were not reabsorbed in the kidney. Bicarbonaturia is easily detected (alkaline urine pH), and ketonuria or anions of drug origin can be suspected from the history (e.g., fasting, aspirin ingestion).

The second group of compounds that can lead to Na loss in the urine with normal kidneys are osmotic diuretics (glucose, urea, mannitol). Alternatively, the ingestion of pharmacologic diuretics also leads to a natriuresis. Failing to find the above in a patient with natriuresis and ECF volume contraction suggest a renal cause for Na wasting. Disease processes by nephron segment would include Fanconi syndrome, Bartter's syndrome, low aldosterone bioactivity, possible atrial natriuretic factor release (supraventricular tachycardias), and renal parenchymal disease (interstitial nephritis, renal tubular necrosis).

K. The urine [K] is evaluated relative to the stimuli for K excretion and Na retention. Since virtually all K is secreted in the cortical distal nephron, evaluate the urine [K] corrected for medullary water abstraction (TTKG, page 224).

TABLE 13–1. Use of Urine Electrolytes

Electrolyte	Normal Response	Patient Response	Potential Pitfalls
Na	• Reflects diet and ECF volume • <10 mM if ECF volume contraction	• >20 mM in ECF volume contraction suggests renal tubular damage	• Diuretic use • Nonreabsorbed anions, e.g., recent vomiting, drugs
K	• Reflects diet, plasma [K], aldosterone action	• If hypokalemia and urine [K] >20 mM, TTKG >7 or rate of K excretion >30 mmol/day, then K excretion too high	• K-sparing diuretics • Low urine [Na] • Water diuresis
Cl	• Reflects diet and ECF volume • <10 mM if ECF volume contracted	• >20 mM with ECF volume contraction suggests renal damage	• Diuretic use • Diarrhea
pH	• Depends on acid-base status • Useful for bicarbonaturia	• Useful once low NH_4^+ excretion confirmed to define cause of low NH_4^+	• Unreliable for urine NH_4^+ • Urinary tract infection
HCO_3	• Depends on diet and acid-base status • >10 mM indicates HCO_3 load • 0 in acidemia	• High urine HCO_3 with chronic metabolic alkalosis indicates vomiting or HCO_3 input • High urine HCO_3 with acidemia in pRTA	• Urinary tract infection • Carbonic anhydrase inhibitors
(Na + K − Cl)	• Depends on diet and acid-base status	• $Na + K > Cl$ = low urine NH_4^+ • $Cl > Na + K$ = high urine NH_4^+	• Ketonuria • Drug anions • Alkaline urine

When assessing the renal contribution to K loss, determine the TTKG and the K excretion rate separately because the K excretion rate is the product of the urine [K] and the urine volume. Hence, polyuric states in which ADH is present can lead to a large renal K loss, owing to the large volume of urine (diabetes mellitus in poor control). Recall that, on a usual diet, most people absorb close to 1 mmol K/kg of body weight and must excrete this quantity of K in their urine to stay in K balance.

Cl. Na and Cl are usually excreted in parallel; hence, Cl excretion also reflects dietary intake, and there is no normal range. During ECF volume contraction, the urine should have an extremely low [Cl]. Exceptions to this rule are the presence of diuretics (osmotic and pharmacologic) and intrinsic renal diseases (e.g., Bartter's syndrome). Na and Cl are not excreted in parallel if ECF volume contraction is accompanied by hyperchloremic metabolic acidosis (e.g., diarrhea) where there is excretion of NH_4^+ plus Cl. As mentioned previously, with recent vomiting, the urine contains Na but no Cl.

HCO_3. HCO_3 appears in the urine following its ingestion or if the salts of organic anions are metabolized (e.g., citrate, lactate). Again, there are wide fluctuations due to diet, and there is no normal range. Bicarbonaturia is detected by finding a urine pH >7.

NH_4^+. This cation should be very low (<25 mM) if alkali is consumed, whereas it should be >80 mM if the kidney is responding normally to metabolic acidosis. Since the urine [NH_4^+] is not usually measured and the urine pH is an unreliable index of the [NH_4^+] in the urine, we recommend that the urine net charge be used as an indirect index of the [NH_4^+] (see page 54 for more detail). As a rough guide, if the urine [Cl] is > the sum of [Na^+] + [K^+], then the kidney is not the primary cause of the metabolic acidosis because the urine NH_4^+ concentration exceeds 80 mM.

Osmolality. Interpret this parameter relative to the plasma [Na] (the need to excrete water). If hyponatremia is present, the urine osmolality should be <100 mOsm/kg H_2O (urine volume 0.5–1 L/hr), whereas during hypernatremia, the urine osmolality should be 1000 mOsm/kg H_2O or greater (and volume of 0.5 L/day).

TABLE 13–2. Urine Electrolytes During Chronic Metabolic Alkalosis Secondary to Vomiting or Diuretics

Electrolyte	Vomiting	Diuretics
Na	• High if recent vomiting despite ECF volume contraction; otherwise <10 mM	• If ECF volume low (<10 mM), unless recent diuretic use
K	• High relative to blood value with recent vomiting • Urine [K] >7 × plasma [K] • Not critical for diagnosis	• Same as above
Cl	• Always close to zero	• Same as above
HCO_3	• Abundant with recent vomiting	• Zero

ADDITIONAL PROBLEMS

Case of Hyponatremia (Flow Chart 7–A, page 171)

Question (Answer on page 295)

2. What is the cause of hyponatremia, and what components of therapy were responsible for correcting hyponatremia?

Analyze the causes for the hyponatremia in a quantitative fashion by assessing separately the contribution of water gain and Na loss.

A 40 year old 70 kg male has chronic panhypopituitarism treated with maintenance hormone replacement (thyroxine, hydrocortisone, testosterone, and DDAVP, an ADH preparation). He was doing well until he had several bouts of vomiting one week ago. Physical examination was normal, including an assessment of his ECF volume. Laboratory values on admission and 7 hr later are summarized below. In addition, the plasma glucose and urea concentrations and the plasma AG were normal.

Therapy consisted of 1 L of isotonic NaCl plus 80 mmol KCl; urine volume was 0.75 L.

	Plasma (0 hr)	Plasma (7 hr)	Urine (during 0–7 hr)
Na (mM)	119	127	100
K (mM)	2.7	3.2	17
HCO_3 (mM)	24	24	—
Osmolality (mOsm/kg H_2O)	250	265	299

Diagnostic Steps in Hyponatremia

1. Rule out pseudohyponatremia by finding a plasma osmolality <270 (correcting for urea and ethanol).

2. If the ECF volume is low, look at urine Na and Cl to determine if there is a renal or non-renal cause for Na loss (Table 13–1, page 291).

3. If the ECF volume is normal, SISADH is the cause (Table 7–3, page 179).

Answer

2. Following the approach outlined in Flow Chart 7–A, page 171, the low plasma osmolality, normal ECF volume, and the absence of hypoosmolar urine suggest that a high ADH is causing hyponatremia, and the patient was receiving an ADH analogue. However, hypokalemia is not an expected feature of excess ADH syndrome (SISADH); it was due to the vomiting (see answer 5, page 237). Could the K loss have influenced the hyponatremia?

By examining the response to therapy, the relative importance of water retention due to DDAVP and a Na shift out of the ECF can be evaluated. The patient received 154 mmol of Na (1 L of isotonic NaCl) and excreted 75 mmol of Na (0.75 L × 100 mM). Thus, he had a 79 mmol positive Na balance. Given a rise in plasma [Na] (Na:H_2O) of 8 mM during therapy and an ECF volume of 14 L, the calculated gain of Na in the ECF is 112 mmol (8 × 14). Thus, there must be an appreciable loss of body water or another source of ECF Na to explain the increased [Na].

The change in water balance was very small, positive by 0.5 L; hence, Na gain in the ECF accounts for the rise in the plasma [Na].

Since the plasma [K] rose 0.5 mM, then 7 mmol of the administered K was retained in the ECF (0.5 mM × 14 L). Furthermore, he excreted close to 13 mmol of K in the urine (17 mM × 0.75 L). Therefore, 80 − 7 − 13 or 60 mmol of K were retained in the body but not in the ECF. Since there was no change in the plasma [HCO_3] after KCl administration, it appears that 60 mmol of K entered the ICF in exchange for an equivalent quantity of Na rather than H^+.

From the above, the final net gain in ECF Na was 79 mmol of infused Na plus 60 mmol of Na that shifted from the ICF to the ECF. Ignoring the water shift induced by raising the plasma [Na] for the moment, the new plasma [Na] would be (119 × 14) + (79 + 60) mmol/14 L, or 129 mM. Now, taking into consideration the volume of water infused and the volume of water shifted from the ICF to the ECF as a result of the ECF Na gain, the final [Na] should be 126 mM, close to the measured 127 mM.

Final Diagnosis. HYPONATREMIA WAS DUE TO BOTH WATER RETENTION (DDAVP ACTION) AND TO A Na SHIFT FROM THE ECF TO THE ICF, OWING TO K DEFICIENCY SECONDARY TO VOMITING.

CORRECTION OF HYPONATREMIA WAS DUE TO POSITIVE EXTERNAL Na BALANCE AND A Na SHIFT FROM ICF TO ECF WITH K RETENTION.

Case of Hypernatremia

Question (Answer on page 297)

3. *What is the cause of hypernatremia in this case?*

A 47 year old male was confused upon admission, so a detailed medical history was not available. Physical exam revealed a normal ECF volume and a disturbed sensorium but no localized findings. Hence, thirst could not be evaluated. The major laboratory abnormality was a plasma [Na] of 158 mM. While in the emergency room, he passed 300 ml of urine/hr and the urine osmolality was 150 mOsm/kg H_2O. There was no change in urine volume or osmolality after ADH was given. The approach to the differential diagnosis of hypernatremia is shown in Flow Chart 8–A, page 195.

Diagnostic Steps in Hypernatremia

1. Rule out Na gain by finding a normal or contracted ECF volume.
2. Rule out a water shift into the ICF if weight loss is present.
3. Rule out non-renal water loss by finding a urine flow rate that is >20 ml/hr or a urine osmolality that is <900 mOsm/kg H_2O.
4. If the urine osmolality is very low, diabetes insipidus is present. Look for ADH response to separate central from nephrogenic causes.
5. If the urine osmolality is in the 350–600 mOsm/kg H_2O range, there is an osmotic or electrolyte diuresis.

Answer

3. The normal ECF volume rules out Na gain as the primary cause of hypernatremia; therefore, a primary cause for water loss must be identified. Since the urine volume was very large and the osmolality very low, diabetes insipidus was present. The failure to respond to ADH makes the diagnosis *nephrogenic diabetes insipidus*. A nephron site involved is the cortical collecting duct. The thick ascending limb of the loop of Henle is not involved because the urine osmolality is distinctly low (it is not the type of nephrogenic diabetes insipidus that just eliminates renal medullary hypertonicity, as an isoosmolar urine would be excreted in that case). Hence, an ADH antagonist should be sought because *partial central diabetes insipidus* was ruled out by the failure to achieve at least an isosmolar urine after ADH was given.

The best bet for an ADH antagonist is lithium. Examination of hospital records revealed that this patient was taking lithium for a manic-depressive psychosis; thus, the final diagnosis was *lithium-induced nephrogenic diabetes insipidus*.

Treatment can be divided into 3 phases:

a. Stop lithium intake; nevertheless, it may take 2–3 weeks to excrete all the retained lithium, and the disorder can occur with therapeutic levels for the plasma lithium concentration.

b. The patient needs a large daily free-water load (at least 4–5 L plus ongoing losses). Since the urine osmolality was 150 mOsm/kg H_2O, the urine [Na] cannot be greater than the infused [Na] of 75 mM with half-normal saline. Hence, this form of therapy is not effective in replacing the free-water loss. Furthermore, since you cannot infuse glucose in water at a rate exceeding 0.3 L/hr without a significant danger of severe hyperglycemia (see page 205 for more detail), you must give water by the GI tract. To do this, give water via an N-G tube, even though the patient is confused. Once the confusion clears, the thirst response returns and the patient should drink enough water to reverse the hypernatremia. It is important to recognize that the confusion and failure to drink led to the rapid deterioration in this case.

c. Lessen the lithium load in the cortical collecting duct cells. Since lithium enters these cells from the lumen via the Na^+ channel, inhibition of this flux with amiloride has been shown to diminish urine volume and raise the urine osmolality.

Suggested Reading

Battle D, von Riotte AB, Gaviria M, et al: Amelioration of polyuria by amiloride in patients receiving long-term lithium therapy. New Engl J Med 312:408, 1985.

Case of Hyperkalemia

Question (Answer on page 299)

4. What is the cause of hyperkalemia in this patient?

A 50 year old male has chronic pancreatitis requiring exocrine enzyme replacement. He had no past history of renal disease, and there was no evidence of insulinopenia. The patient was admitted to the hospital for a 3-week period of total parenteral nutrition for "pancreatic rest." No medication was given. The first plasma [K] before this regimen was 5.7 mM. Hyperkalemia was a problem throughout the hospital course. Physical exam was normal except for the evidence of muscle wasting. Specifically, there was no evidence of ECF volume contraction or skin pigmentation.

On laboratory exam, there was no evidence of hemolysis, thrombocytosis, or circulating tumor cells. Serum enzymes were normal, ruling out tissue necrosis. Specific data are shown below. In addition to the values listed, the following were normal: plasma creatinine of 0.7 mg/dl (64 uM); serum [Na], [Cl], and [HCO_3] of 141 mM, 102 mM, and 23 mM respectively, and the plasma glucose concentration was 72 mg/dl (4 mM). The urine electrolytes on day 14 were as follows: Na = 61 mM, K = 7 mM, Cl = 62 mM.

Plasma [K] and K Infusion Rate on TPN

Day	Plasma [K] (mM)	K Infusion (mmol/day)	K Exalate Resin (g/day)
1	5.7	70	0
4	4.2	30	60
6	4.1	30	0
8	5.3	15	0
12	6.5	15	60
14	5.2	0	0

Urine Electrolytes on 3 Hospital Days

Day	Plasma [K] (mM)	Urine (mM or mOsm/kg H_2O)			
		Na	*K*	*Cl*	*Osm*
*14**	5.2	61	7	62	285
16	5.6	17	74	—	357
24	5.2	45	7	48	—

*On day 14, a specific diagnostic test was performed. What was it and what were the results?

Answer

4. The first step in the differential diagnosis of hyperkalemia is to be sure that the [K] is a real value. Given the absence of hemolysis, thrombocytosis, and circulating tumor cells, the hyperkalemia appears to be real rather than a laboratory problem.

Next, evaluate K intake as a cause, albeit unlikely, of hyperkalemia. K intake by diet was nil, and via TPN was zero even on days 13 and 14. Hence, excessive K intake was not the problem.

Next, determine whether K loss was abnormally low. Since the K excretion rate, urine [K], and TTKG were all excessively low, reduced K excretion looks like the primary cause for hyperkalemia (urine K excretion = urine volume × urine [K]). Since the urine volume was 1.5 L/day, the problem appears to be with the urine [K]. To raise this [K], Na must be delivered to the cortical distal nephron, and aldosterone must act on this nephron segment. Since the urine [Na] was 61 mM on day 14, a low Na delivery was not a cause of the problem. Hyperkalemia should have stimulated aldosterone release; hence, the patient appears to have an adrenal cortical problem or a renal disease, resulting in low response to aldosterone. These two possibilities could be differentiated by administering a physiologic dose of 9 α-fluorohydrocortisone. However, one test was done first:

Patients with aldosterone deficiency excrete their dietary K load (60 mmol) in a volume of approximately 1.5 L of urine; hence, their urine [K] is close to 40 mM. The finding of a much lower rate of K excretion in this patient suggests an alternative problem to low aldosterone bioactivity. Most commonly, such a low urine [K] would suggest the presence of K-sparing diuretics; however, these were not taken. Since hyperkalemia was not present on days 4–6 and there was some K loss via the urine and the resin over several days with no K intake, the plasma [K] should have decreased. Hence, the plasma [K] should be reevaluated.

To determine if there was an abnormal release of K from muscle, owing to the blood sampling procedure, blood was drawn in the conventional way, using a tourniquet, while a free-flow sample was obtained from the femoral vein. Simultaneous values revealed a plasma [K] of 5.7 mM from the arm and 3.4 mM from the femoral vein. Therefore, there was spurious hyperkalemia due to the sampling procedure. Although the final diagnosis is not absolutely clear, either muscle contraction with blood drawing led to local hyperkalemia and washout (augmented by muscle wasting) or there was a specific defect in muscle energy metabolism related to the patient's poor nutritional state.

Although it might seem academic to work this problem to completion, it is important to do so. Since the patient is actually hypokalemic while on TPN, the synthesis of new muscle or the development of diarrhea (with K loss) could lead to much more severe hypokalemia.

Case of Hyperkalemia

Question (Answer on page 301)

5. What is the cause of hyperkalemia in this case? Is it due to low aldosterone levels?

A 73 year old woman with carcinoma of the breast and bone metastases was admitted to the hospital for chemotherapy. Her only medication was an amiloride-containing diuretic for chronic hypertension. Physical examination and laboratory results were all normal, except that her serum creatinine (1.4 mg/dl, 123 uM) and BUN (35 mg/dl; urea, 12 mM) were slightly elevated. Chemotherapy included aminoglutethimide (250 mg q i d) and hydrocortisone (37.5 mg/day). Aminoglutethimide inhibits adrenal steroid hormone (including aldosterone) synthesis. After 2 weeks of therapy, the patient complained of marked weakness, and the plasma electrolytes were obtained. Physical exam revealed marked ECF volume contraction. The urine volume was small, and the urine electrolytes are shown below. At no time did the patient suffer from hypercalcemia or hyperglycemia. EKG revealed abnormalities consistent with hyperkalemia.

Treatment consisted of 2 L of isotonic NaCl to reexpand the ECF volume, 150 mmol of $NaHCO_3$, and 50 g of glucose to shift K into cells. She was also given intravenous Ca^{++} to antagonize the effects of hyperkalemia and Kayexalate by both rectal and oral routes to promote K loss from the gastrointestinal tract.

Urine and Plasma Data

Parameter	Plasma	Urine
Na (mM)	138	139
K (mM)	9.3*	30
Cl (mM)	111	121
HCO_3 (mM)	19	0
pH	7.35	5.5
Osmolality (mOsm/kg H_2O)	300	658
Volume	—	"Low"
Aldosterone	"High normal"	—
Renin	"High"	—

*The plasma [K] fell to 6.2 mM at 6 hr and 5.5 mM at 9 hr.

Answer

5. The first step in the differential diagnosis of hyperkalemia is to rule out a laboratory error. There was no evidence of blood cell lysis.

The next step is to establish whether the kidney is playing an important etiologic role in the hyperkalemia. Since the urine [K] was very low, there was an inappropriate response to hyperkalemia. There are two major possibilities to consider: low delivery of Na to the cortical collecting duct, owing to ECF volume contraction, and the failure to respond to aldosterone because of its absence due to aminoglutethimide or to a renal problem (amiloride, in this case). The first possibility was ruled out by the high urine [Na] (due to the diuretic action).

We could suspect that amiloride was acting because hyperkalemia per se should cause an elevated urine [K] even in the absence of aldosterone. At this point, let us look at the circulating aldosterone level—it is at the high end of the usual normal range and should be high enough to cause an elevation in the urine [K], again pointing to an important role for amiloride in the etiology of hyperkalemia.

Parenthetically, note that despite the two major stimuli for aldosterone release (hyperkalemia, high renin), the aldosterone level is much lower than you might expect but still high enough to elicit some renal response. This relative reduction in aldosterone could reflect the action of aminoglutethimide.

Amiloride, the drug that inhibits the Na^+ channel in the luminal membrane, has a very long half-life, perhaps 12 hr or longer in this patient with marked ECF volume contraction; hence, the only leverage to increase K excretion is to increase urine volume (K excretion = urine [K] × urine volume).

Urine volume rises with saline administration; a loop diuretic should be added even though the ECF volume is reduced, providing that the saline infusion rate exceeds the natriuresis and diuresis. In this case, the ECF volume will increase, and K excretion can be increased promptly.

Within 12 hr after discontinuing amiloride, there was a progressive rise in the urine [K]. However, since the plasma aldosterone levels are not known at the bedside, the patient should also be given the mineralocorticoid 9 α-fluorohydrocortisone in addition to the other means of therapy as long as hyperkalemia persists.

With respect to a K shift from cells, the patient did not have hormonal deficits, necrosis, or metabolic acidosis.

Case of Metabolic Acidosis

Questions (Answers on page 303)

6. A 21 year old female has had diabetes mellitus for 2 years, requiring insulin treatment. She presented with lethargy, malaise and headaches. She had metabolic acidosis with a normal plasma AG. Her complaints and the acid-base disturbance persisted for 6 months, so she was referred for further evaluation. She denied diarrhea or abdominal complaints. She had ingested no drugs (Diamox, halides, or HCl equivalents). She had no past history of renal disease.

While taking 34 units of insulin lente daily, she frequently had glycosuria and ketonuria, but no major increase in the plasma AG. Her physical examination was unremarkable, and her urinalysis showed no protein and a normal sediment. The results of her laboratory investigations are shown below:

	Plasma		
BUN (mg/dl [mM])	11 (3.9)	Na (mM)	136
Creatinine (mg/dl [mM])	0.9 (100)	K (mM)	2.9
Glucose (mg/dl [mM])	190 (10.6)	Cl (mM)	103
pH	7.35	HCO_3 (mM)	19
$[H^+]$ (nM)	45	Plasma AG (mEq/L)	14
P_{CO_2} (mm Hg)	35	β-HB (mM)	2.2
	Urine		
Glucose (mM)	5	Na (mM)	47
Urea (mM)	50	K (mM)	60
pH	5.3	Cl (mM)	93
Osmolality (mOsm/kg H_2O)	680		

BHB = β-hydroxybutyrate

Questions (Answers on page 303)

6A. What is your differential diagnosis?

6B. What investigative plan would you organize?

Additional Questions (Answers on page 303)

Following the ingestion of $NaHCO_3$ and KCl, the plasma $[HCO_3^-]$ rose to 26 mM and did not fall acutely; the urine P_{CO_2} was 90 mm Hg.

6C. What is your diagnosis now?

6D. Why was this patient hypokalemic?

Answer

6A. The patient had hypokalemia and metabolic acidosis, suggesting proximal RTA or reduced distal H^+ secretion. The urine net charge suggested that there was little NH_4^+ excretion, supporting a diagnosis of distal RTA (patients with proximal RTA alone should have high NH_4^+ excretion rates when the urine is free of HCO_3 [page 87]). The differential diagnosis, at this point, would be the causes of low distal H^+ secretion associated with hypokalemia (page 92).

6B. The plan would be to confirm the absence of proximal RTA with HCO_3^- administration and then to confirm low distal H^+ secretion by measurement of the urine Pco_2.

6C. The HCO_3 administration indeed ruled out proximal RTA. However, the urine Pco_2 was normal, indicating either H^+ back-diffusion from lumen to cell when the urine pH was low (H^+ backleak type of distal RTA), or else there was a normal distal H^+ secretory process and a more complex disease. As the patient had not used amphotericin B or toluene, backleak of H^+ in the collecting duct is an unlikely diagnosis. The other possibility was that we were being misled by the urine "apparent net charge"—that there indeed was high NH_4^+ excretion but that it was in conjunction with an anion other than Cl. This possibility can be confirmed by measuring the urine osmolality and comparing it with the calculated urine osmolality (see page 97).

The osmolality of the urine was 680 mOsm/kg H_2O, whereas the calculated osmolality is the urine [Na] + [K] + [Cl] + glucose + urea = 269 mOsm/kg H_2O. Therefore, there is indeed an "osmolal gap" of 411 mOsm/kg H_2O, indicating the presence of a large number of unmeasured osmoles (possibly NH_4^+ + β-hydroxybutyrate). The urine NH_4^+ and β-hydroxybutyrate were measured and were 220 mM and 234 mM, respectively. Since the urine volume was >1 L/day, the patient had a normal renal response to acidemia, excreting close to 300 mmol of NH_4^+ per day and had acidemia caused by an ongoing acid load due to ketoacidosis; this was not evident because of marked ketonuria (see Fig. 3–1, page 51). With better diabetic control, the acidemia disappeared.

6D. In the face of hypokalemia, the patient had a urine [K] of 60 mM. Back-correcting for medullary water abstraction, her [K] in the cortical collecting duct was close to 30 mM (60 mM/680/plasma Osm), and this was about 10-fold greater than her plasma [K]. Thus, aldosterone was acting on the kidney, and aldosterone was released in response to ECF volume contraction due to renal Na loss secondary to the ketonuria. Thus, she has hypokalemia secondary to renal K loss. Furthermore, despite insulin deficiency, the plasma [K] was 2.9 mM; this suggests a very large K deficit.

Hypokalemia in a Severely Alkalemic Patient
Illustrative Case

7. A 23 year old male has been on an alcoholic binge associated with numerous bouts of vomiting. He was brought to the hospital in a confused state. He was afebrile on admission but appeared very ill. On physical examination, he was hyperventilating; blood pressure was 70/50, pulse 120 beats/min. No other positive findings were noted.

Laboratory results are summarized below. Of special note was the severe alkalemia and hypokalemia.

Parameter	Plasma	Urine
Na (mM)	125	65
K (mM)	2.3	45
Cl (mM)	54	0
HCO_3 (mM)	36	—
pH	7.70	8.0
P_{CO_2} (mm Hg)	30	—
P_{O_2} (mm Hg)	80	—
Osmolality (mOsm/kg H_2O)	290	600

Questions (Answers on page 305)

7A. Why is this patient so hypokalemic? What additional concerns does the hypokalemia raise with respect to therapy? What is your therapeutic regimen with respect to K?

7B. Why is this patient so alkalemic? What changes might occur in pH early during therapy? What is your recommended acid-base therapy?

Answers

7A. Severe hypokalemia is the result of K loss and a K shift into cells. The former is probably more important and occurred primarily via urine K loss. The urine K loss was caused by aldosterone (released because of ECF volume contraction) action together with high distal Na delivery (secondary to the bicarbonaturia due to recent vomiting; see Fig. 4–1, page 99).

The degree of K shift cannot be assessed with any certainty in this case. K could have shifted into cells, owing to the alkalemia of the metabolic alkalosis. However, the insulin deficiency (secondary to α-adrenergic action on B cells) would have raised the plasma [K].

Therapy with saline to increase the ECF volume toward normal should increase urine volume along with increased $NaHCO_3$ excretion. Therefore, the rate of K excretion should increase, owing to the increased distal Na and volume delivery in a patient with high aldosterone levels. Hence, this patient might develop acute severe K depletion because K loss could transiently exceed K input. Therefore, it would be prudent to give 1 dose of a "K-sparing" diuretic (amiloride). If you do, watch the plasma [K] and adjust the rate of K administration accordingly. In summary, the therapeutic regimen is saline for ECF volume depletion, KCl to replace the K deficit, and a K-sparing diuretic to diminish anticipated additional early K loss.

7B. The alkalemia is due to both respiratory (low P_{CO_2}) and metabolic alkalosis (high plasma [HCO_3]) during alkalemia. The respiratory alkalosis was most likely due to aspiration pneumonitis or a pulmonary embolus (note the large A—a gradient of 25 mm Hg); the metabolic alkalosis was due to vomiting.

Early in therapy, oxygen delivery to tissues increases when the ECF volume increases. Note that the plasma AG is markedly elevated (12 + 23 mEq/L), largely reflecting circulating lactate and β-hydroxybutyrate (due to the low oxygen delivery, ethanol metabolism, and α-adrenergic inhibition of insulin release from β cells). Since these anions are metabolized rapidly (consuming H^+), you should expect a rise in the plasma [HCO_3]; this rise will be somewhat less than expected, owing to the infusion of a HCO_3-free solution and some bicarbonaturia. This rise in [HCO_3] could worsen the alkalemia and, in a hypokalemic patient, may precipitate a cardiac arrhythmia.

The therapy for metabolic alkalosis (NaCl, KCl) might initially aggravate the alkalemia. Consider whether the patient would be better if measures were taken to raise the P_{CO_2} or to lower the plasma [HCO_3] (infuse HCl or NH_4Cl). These issues are controversial but are raised to show that treatment needs to be individualized.

14

EMERGENCY APPROACH TO PATIENTS WITH ACID-BASE, FLUID, AND ELECTROLYTE DISORDERS

GENERAL INTRODUCTION

This chapter was designed to help the physician deal with a fluid, electrolyte, or acid-base emergency. Only the "bare bones" are provided. Critical physiology, diagnostic features, investigations, and therapeutic recommendations are highlighted. Guidelines, which include normal values and expected physiological responses, are provided.

We have posed several leading questions that the physician should ask, and we provide the answers to these questions, which help to formulate an approach to the problem.

We have also reproduced the flow sheets to help summarize the steps to be taken to reach a final diagnostic group. At a later time, the physician should read the clinical and physiology chapters for more comprehensive information.

METABOLIC ACIDOSIS

Definition

Blood [H] > 45 nM and HCO_3 < 24 mM. (See page 64 for wide anion gap variety and page 86 for normal anion gap variety.)

Expected Normal Response

Lung: For every 1 mM reduction in [HCO_3] from 25, the

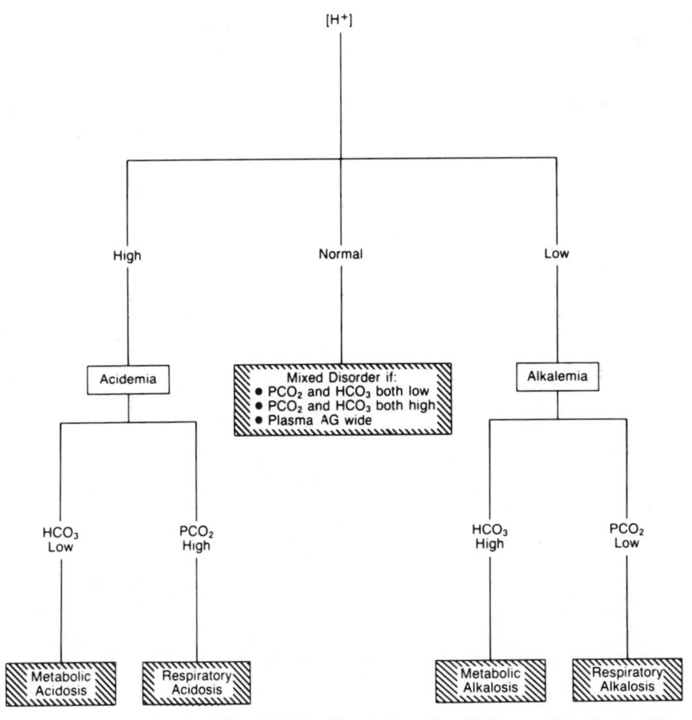

Flow Chart 14–A: The initial diagnosis of acid-base disorders. The information to be interpreted to make a final diagnosis is indicated in the clear boxes. The final diagnoses are shown in the shaded boxes.

P_{CO_2} should fall 1 mm Hg from 40.

Kidney: The kidney should increase NH_4^+ excretion to improve renal new HCO_3 generation (> 100 mmol/day).

Diagnostic Clues

Cause	Distinguishing Features
Organic acid production	• Wide anion gap • Rapidly progressive
Renal failure	• Wide anion gap • Slowly progressive
Renal tubular acidosis	• Normal anion gap
Distal	• Slowly progressive • Low urine $[NH_4]$ (urine [Cl] < [Na] + [K])
Proximal	• Urine pH > 7 when plasma $[HCO_3]$ approaches 25 mM
GI HCO_3 loss	• Normal plasma gap • May progress rapidly • High urine NH_4 (urine [Cl] > [Na] + [K])

Clinical Pearls

If plasma $[HCO_3]$ < 8 mM, give HCO_3; calculate amount by distribution in volume equal to 50% of body weight.

If [K] < 4 mM, use $KHCO_3$ plus some $NaHCO_3$.

If severe acidemia and respiratory acidosis, ventilate.

Patients with normal plasma anion gap require more exogenous HCO_3 than those with wide anion gap.

Hypoalbuminemia decreases the value for plasma AG.

If the plasma anion gap is normal and the cause not evident, suspect inapparent organic acid production.

Rules to Memorize

Anion gap = [Na] − [Cl] − $[HCO_3]$ = 12 ± 2 mEq/L.

For every mM fall in $[HCO_3]$ from 25 mM, expect the P_{CO_2} to fall by 1 mm Hg from 40 mm Hg.

Essentials of Treatment

If $[HCO_3]$ < 8 mM, raise to 12 mM with HCO_3.

Give HCO_3 earlier if plasma anion gap is normal.

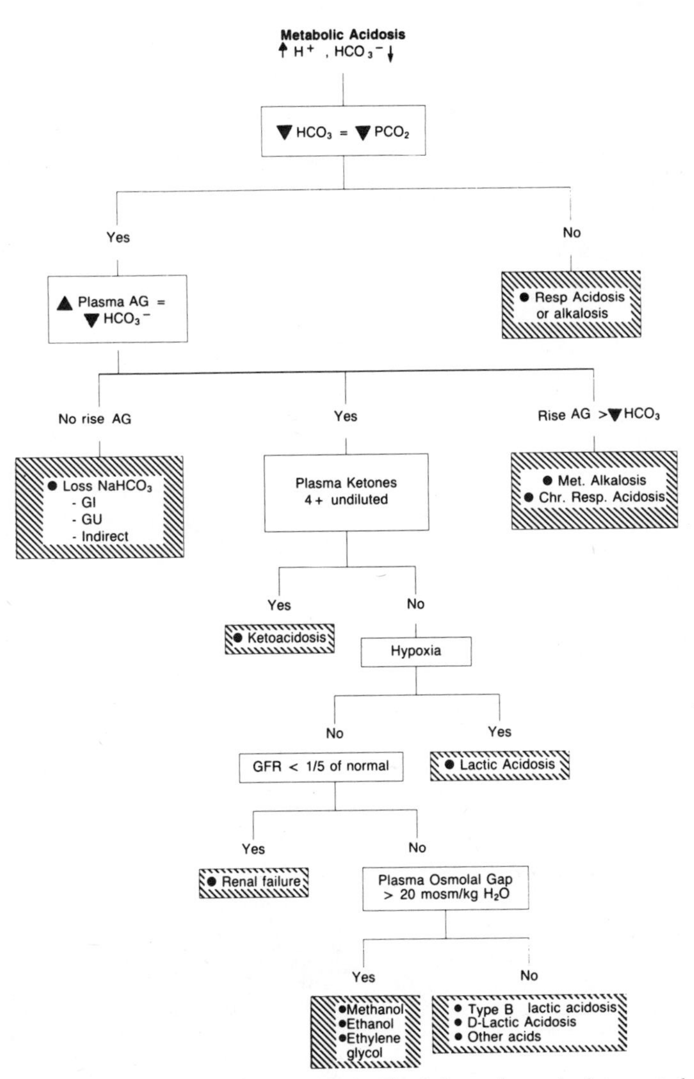

Flow Chart 14–B: Metabolic acidosis. The information to be interpreted to make a final diagnosis is indicated in the clear boxes. The final diagnoses are shown in the shaded boxes.

WIDE ANION GAP METABOLIC ACIDOSIS

Definition

Plasma anion gap > 15 mEq/L.

Expected Normal Response

In organic acid acidosis, the fall in [HCO_3] should equal the rise in the plasma anion gap.

Diagnostic Clues

Cause	Distinguishing Features
Ketoacidosis	
Diabetic ketoacidosis	• ECF volume contraction • Hyperglycemia • Acetone on breath • Plasma ketones > 1:8 • Usually hyperkalemia
Alcoholic ketoacidosis	• ECF volume very low • Acidemia may be mild or absent • Plasma ketones less positive • Plasma [K] may be low
Lactic Acidosis	
L-Lactic acidosis	• Generally in shock • Severe tissue hypoxia • Elevated blood [lactate]
D-Lactic acidosis	• GI stasis or antibiotics
Renal failure	• Serum creatinine > 5 × normal
Intoxications	
Methanol	• CNS complaints, especially blurred vision • Fruity odor on breath • Wide serum osmolal gap • Methyl alcohol in serum
Ethylene glycol	• Wide serum osmolal gap • Oxalate crystals in urine • Renal failure • Ethylene glycol in serum
Salicylates	• Hyperventilation and tinnitus • Elevated serum salicylate • Look for other causes if acidosis severe

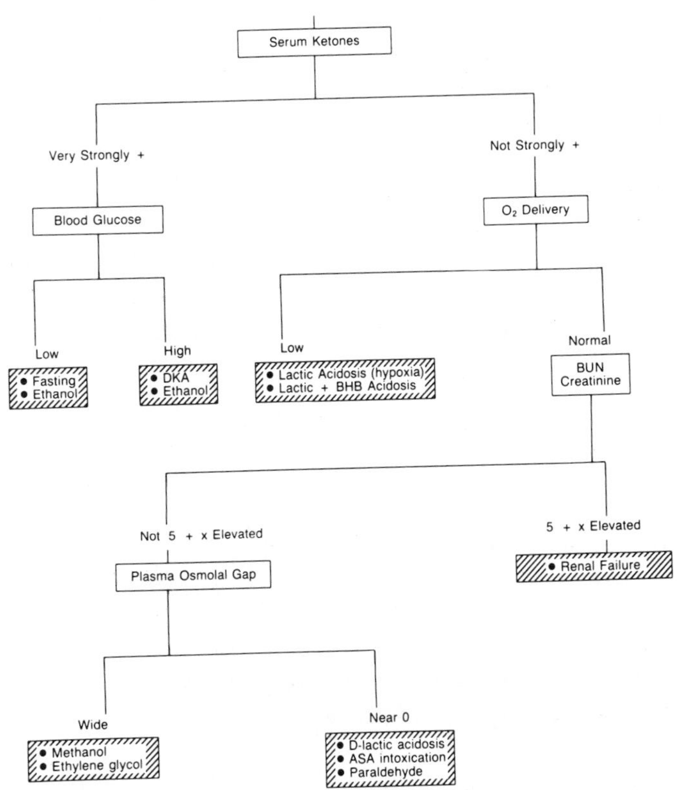

Flow Chart 14–C: Approach to metabolic acidosis with a wide plasma anion gap. The information to be interpreted to make a final diagnosis is shown in the open boxes; the final diagnoses are shown in the shaded boxes. The first critical question is, "Is the plasma unmeasured anion gap wide?"

Clinical Pearls

If fall in $[HCO_3]$ < increase in plasma anion gap, suspect coincident metabolic alkalosis.

If fall in $[HCO_3]$ > increase in plasma anion gap, suspect component of nonanion gap metabolic acidosis.

Use the serial fall in plasma anion gap to monitor the course of the metabolic basis of the acidosis.

Most ketoacidosis will have residual nonanion gap metabolic acidosis after therapy, owing to ketonuria.

Most patients with ketoacidosis are K-depleted, despite initial hyperkalemia; therefore, anticipate hypokalemia with therapy.

Essentials of Treatment

Diabetic Ketoacidosis

Regular insulin, 5–10 unit bolus and 2–10 units/hr IV.

Na deficit, 5–10 mmol/Kg:

Give isotonic saline until hemodynamically stable, then 75 mM saline.

Add 20–40 mM KCl to saline infusions.

Give HCO_3 if $[HCO_3] < 8$ mM.

L-Lactic Acidosis

Restore tissue O_2 delivery by improving BP, cardiac output, and PO_2 if low.

Give HCO_3 if $[HCO_3] < 8$ mM.

Methyl Alcohol and Ethylene Glycol

If high index of suspicion, begin the following therapy prior to confirmation of diagnosis:

QUESTIONS TO ASK WITH METABOLIC ACIDOSIS

1. **Is the respiratory response normal?**
Answer: Compare the fall in P_{CO_2} from 40 to the fall in $[HCO_3]$ from 25; they should be equal.

2. **Is there a reason for the plasma anion gap to be lower than normal?**
Answer: Look for hypoalbuminemia; an error in Na, Cl, or HCO_3; halide ingestion; or dysproteinemia. Save a sample of plasma and urine for later analyses.

3. **What should be done if the plasma anion gap is wide?**
Answer: Suspect the cause clinically and confirm with lab studies (serum ketones or [BHB] for ketoacidosis, plasma creatinine for renal failure (at least 5 $\times$ normal), plasma L- or D-lactate if appropriate, plasma osmolal gap for methanol or ethylene glycol intoxications, and toxicology screen).
If the plasma osmolal gap is wide and you cannot detect ethanol on the breath, give the patient 4 oz whiskey po or 0.6 g ethanol/kg IV to prevent damage due to methanol or ethylene glycol metabolism.
Compare the rise in plasma anion gap to the fall in plasma $[HCO_3]$ to determine if there is a mixed type of acid-base disorder.

4. **What do I do if the plasma anion gap is normal?**
Answer: Establish the renal role in acidemia by estimating the urine NH_4 from the urine *net charge* or anion gap; if urine [Cl] is $>$ [Na] + [K], the most likely cause is GI $NaHCO_3$ loss.
If there is a renal cause, the plasma [K] is very helpful; low values suggest a H^+ secretion defect (confirm with urine P_{CO_2}), whereas high values suggest a problem with aldosterone action (also called Type IV RTA).

5. **When should $NaHCO_3$ therapy be started?**
Answer: If the plasma $[HCO_3]$ is $<$ 8 mM and the plasma [K] is not $<$ 3 mM, give HCO_3 to raise the plasma $[HCO_3]$ to 12 mM (based on a HCO_3 space = 50% of body weight; more $NaHCO_3$ is required when the plasma $[HCO_3]$ is very low). Other situations in which you might use $NaHCO_3$ are for the patient
   a. who cannot make HCO_3 quickly (i.e., by metabolism of high [lactate] or [BHB]).
   b. with very low pH (acidemia compromises heart function).
   c. with kidneys that cannot make appreciable HCO_3.

Oral: 4 oz whiskey initially and 2 oz hourly.
IV: 0.6 gm/kg initially and 0.15 gm/kg/hr in chronic drinkers and half this rate in nondrinkers. You are aiming for a blood ethanol level of 100 mg% (22 mM).
Treat acidemia aggressively with $NaHCO_3$.
If acute renal failure, beware of pulmonary edema.
Plan urgent dialysis if methanol > 50 mg/dl (15 mM).

NORMAL ANION GAP METABOLIC ACIDOSIS

Diagnostic Clues

Cause	Distinguishing Clinical Features
GI HCO_3 loss	• Urine NH_4 > 100 mmol/day • Urine [Cl] >> ([Na] + [K]) • Hypokalemia common
Proximal RTA	• Urine pH alkaline, with serum [HCO_3] below normal • Alkaline urine PCO_2 > 70 mm Hg
Distal RTA	• Urine NH_4 <50 mmol/day • Urine ([Na] + [K]) > [Cl] • Alkaline urine PCO_2 close to blood PCO_2

Clinical Pearls

RTA is acidemia with reduced urine NH_4 excretion.

Hyperkalemia suppresses ammoniagenesis and renal NH_4 excretion.

Urine pH is a poor reflection of renal H^+ secretion, but a urine pH > 7.0 indicates significant bicarbonaturia.

Acidemia due to distal RTA is slowly progressive.

Hypokalemia and severe acidemia (GI or distal RTA) requires special care in therapy (see below).

Essentials of Treatment

HCO_3^- causes K^+ entry into cells; therefore, HCO_3 therapy in hypokalemic patients requires aggressive K^+ replacement and cardiac monitoring. Use $KHCO_3$ if available. Slow HCO_3 therapy when [HCO_3] nears normal; watch urine pH to diagnose proximal RTA. Once urine is alkaline, measure urine PCO_2 to assess distal H^+ secretion.

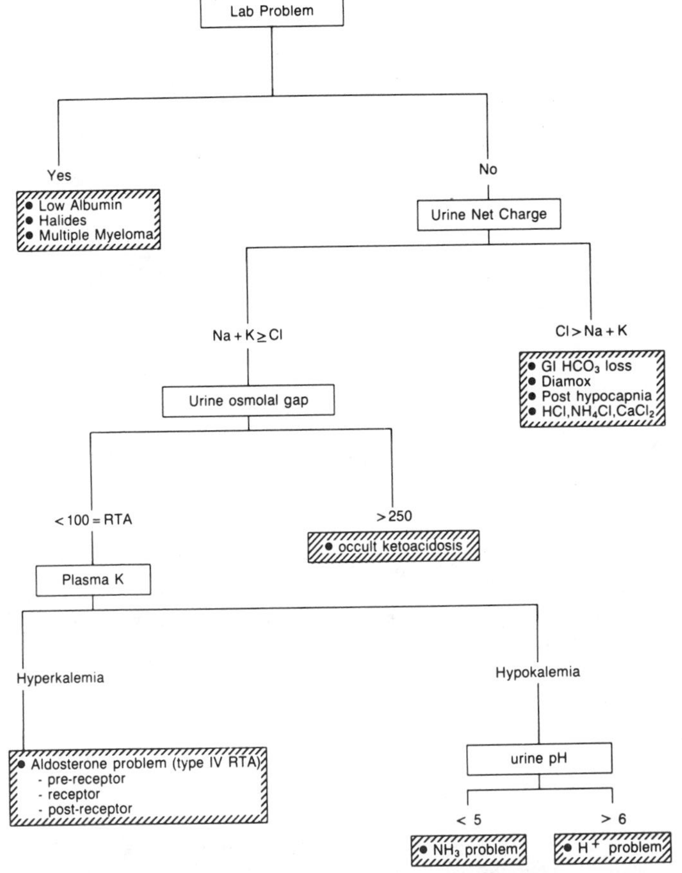

Flow Chart 14–D: Approach to the patient with metabolic acidosis and a normal plasma AG. The information to be interpreted to make a final diagnosis is shown in the open boxes; the final diagnoses are shown in the shaded boxes.

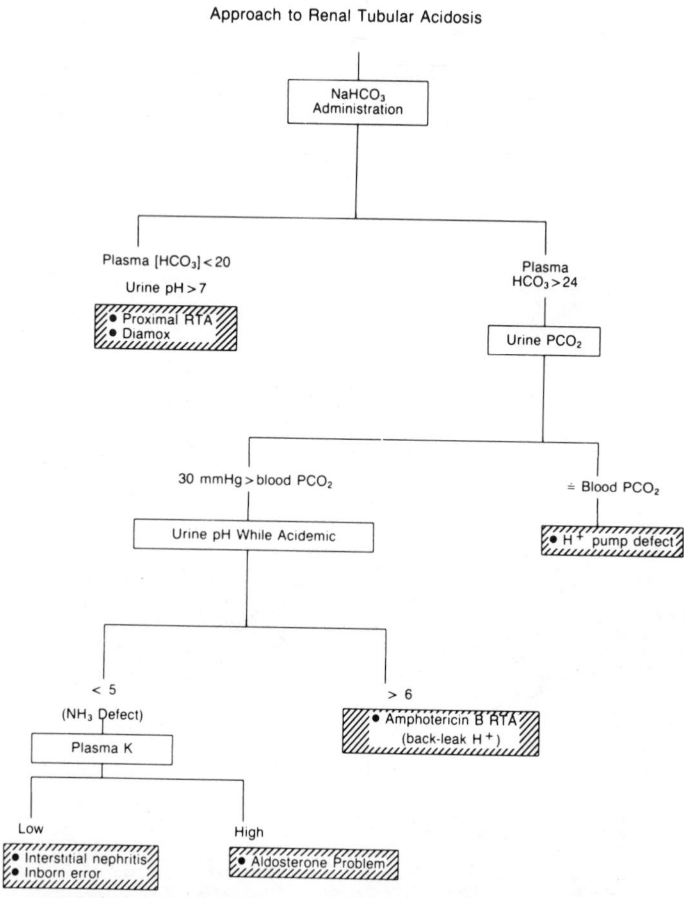

Flow Chart 14–E: Approach to the patient with RTA. The information to be interpreted to make a final diagnosis is shown in the open boxes; the final diagnoses are shown in the shaded boxes.

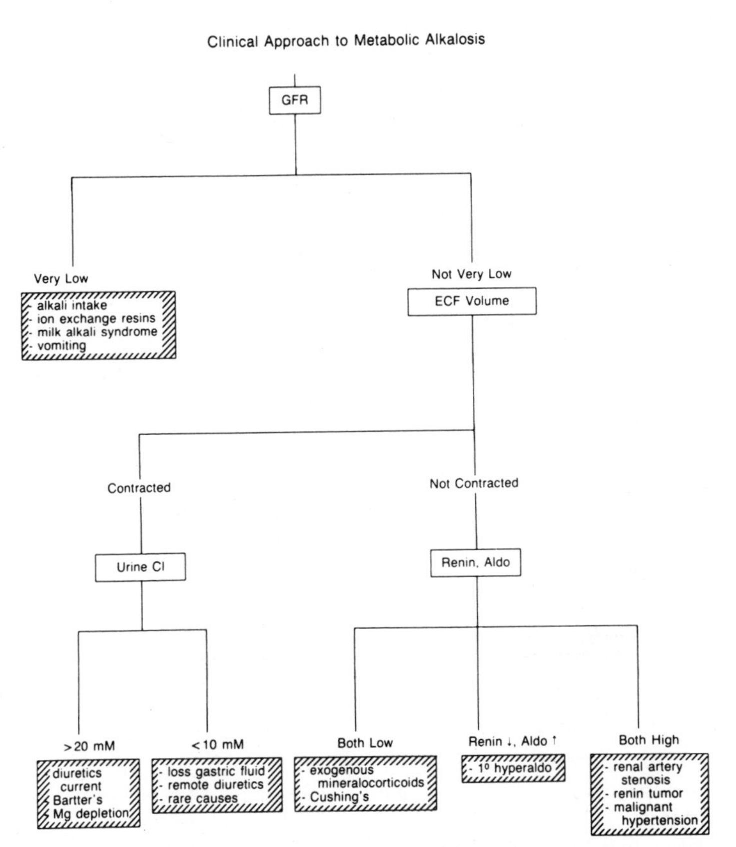

Flow Chart 14–F: Approach to the patient with metabolic alkalosis. The information to be interpreted to make a final diagnosis is shown in the open boxes, whereas the final diagnoses are shown in the shaded boxes.

METABOLIC ALKALOSIS

Definition

Serum [HCO_3] > 25 mM and [H^+] < 38 nM (pH > 7.42).

Expected Normal Response

Lungs: For every 1 mM rise in [HCO_3] from 25 mM, expect a 0.7 mm Hg increase in P_{CO_2} from 40 mm Hg.

Kidney: If ECF volume contracted, the urine should contain little Cl and/or Na (Table 13–1, page 291).

Diagnostic Clues

Cause	Distinguishing Clinical Features
ECF volume contraction	
Vomiting	• Urine Cl < 15 mM • Urine pH > 7 and Na > 20 mM if recent vomiting
Diuretics	• Urine Na and Cl > 20 mM when diuretic acts and < 10 mM other times
Bartter's syndrome	• Urine Na and Cl > 20 mM; no fluctuation
Primary hyperaldosteronism	• No ECF volume depletion • Urine Cl > 20 mM
Mg depletion	• Low serum Mg

Clinical Pearls

Patients might not admit to vomiting or diuretic abuse.

Repetitive urine Na, Cl, and pH tests are the best way to unmask abusers.

Suspect Mg depletion if GI disease, malnutrition, or drugs (aminoglycosides, cisplatin).

Often the hypokalemia is more of a problem than the metabolic alkalosis.

Essentials of Treatment

Correct ECF volume depletion with NaCl; give KCl. Add amiloride if very hypokalemic to prevent further K loss with ECF volume restoration. If hypertensive, can use mineralocorticoid blocking agents.

QUESTIONS TO ASK WITH METABOLIC ALKALOSIS*

1. **Is the ECF volume contracted?**
Answer: A contracted ECF volume helps identify those patients who have lost NaCl and who will respond to replacement of the Na/K/Cl deficits.

2. **Why is the ECF volume contracted?**
Answer: Most commonly, there was diuretic use or vomiting (these may not be admitted). Remember that subtle decreases in ECF volume are very hard to detect clinically; the urine electrolytes in the absence of diuretics can help in this regard.

3. **If ECF volume contraction is present, is the renal response appropriate?**
Answer: The urine should be Cl-free; if not, suspect diuretics, Mg depletion, or Bartter's syndrome (with diuretics, urine Cl loss will be intermittent). A high urine [Na] can occur after recent vomiting.

4. **What tests should be done if the ECF volume is not contracted?**
Answer: Look for the cause of high aldosterone levels (Mg depletion, high renin states, or primary hyperaldosteronism). Hypercalcemia can be a rare cause of this metabolic alkalosis.

5. **How is metabolic alkalosis treated?**
Answer: Metabolic alkalosis is not a specific disease; therefore, treat the cause. In general, NaCl and KCl will correct the *saline-responsive* types.

---

*For the approach to metabolic alkalosis, see Flow Chart 14–F, page 317.

RESPIRATORY ACID-BASE DISORDERS

Definition of Respiratory Acidosis

$P_{CO_2} > 40$ mm Hg and $[H^+] > 40$ nM (pH <7.40).

Also: If $[HCO_3]$ reduced (metabolic acidosis) and P_{CO_2} not appropriately reduced (1:1 with $[HCO_3]$), respiratory acidosis is present.

Definition of Respiratory Alkalosis

$P_{CO_2} < 40$ mm Hg and $[H^+] < 40$ nM.

EXPECTED NORMAL RESPONSE

See "Rules to Memorize."

Diagnostic Clues

Cause	Distinguishing Clinical Features
Respiratory Acidosis	
Will not breathe:	
Due to cerebral nonrhythmic neuronal damage*	• Cannot talk, cough, or hold breath
Due to brainstem rhythmic neuronal damage*	• Abnormal response to O_2 and CO_2
Due to absent upper airway reflexes*	• Cannot swallow; absent nasal, tracheal, and pharyngeal reflexes
Cannot breathe:	
Due to muscle disease†	• Reduced maximum inspiratory and expiratory pressures
Due to increased elastic work†	• Reduced lung volumes • Reduced lung compliance
Due to increased resistance to flow†	• Increased airway resistance

*See Table 5–1, page 129, for details.
†See Table 5–2, page 129, for details.

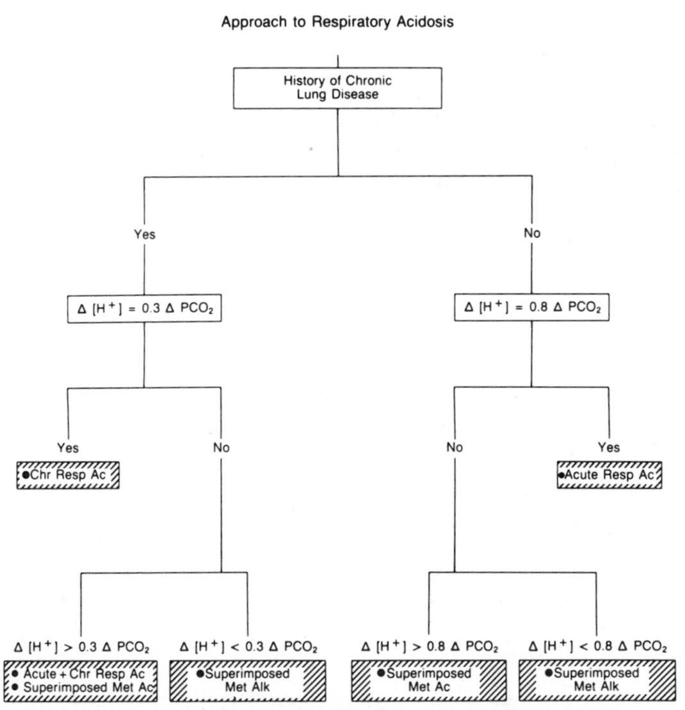

Flow Chart 14–G: Approach to the patient with respiratory acidosis. The variables to evaluate are shown in the open boxes, and the final diagnoses are shown in the shaded boxes.

Diagnostic Approach to Respiratory Acidosis

In a patient with an elevated P_{CO_2}, begin with the *clinical* decision as to the presence of chronic lung disease. If there is no evidence of chronic lung disease, the patient is assumed to have acute respiratory acidosis. If the patient is more acidemic than expected, there is a superimposed metabolic acidosis and, if more alkalemic, a superimposed metabolic alkalosis. On the other hand, if there is a history of chronic lung disease and the patient is more acidemic than predicted, he or she may either have a component of acute respiratory acidosis or metabolic acidosis; if more alkalemic, there is a superimposed metabolic alkalosis.

Clinical Pearls

Whether a respiratory disturbance is acute or chronic is decided on *clinical* grounds, not on the basis of the laboratory results.

Use alveolar-arterial O_2 gradient to distinguish problems of respiratory stimulation from intrinsic lung disease:

$$P_{alv}\ O_2 = P_{insp}\ O_2 - 1.25\ PCO_2$$

In respiratory alkalosis look for serious underlying disease (e.g., sepsis or ASA toxicity).

In patients with chronic respiratory acidosis on diuretics, watch for coincident metabolic alkalosis—it often worsens the clinical state (suspect when plasma $[H^+]$ is normal).

Rules to Memorize

Respiratory Acidosis

Acute: For every 1 mm Hg increase in PCO_2 from 40 mm Hg, expect a 0.77 nM increase in $[H^+]$ from 40 nM; or for every doubling of the PCO_2, expect a 2.5 mM increase in $[HCO_3]$.

Chronic: For every 1 mm Hg increase in PCO_2 from 40 mm Hg, expect a 0.32 nM increase in $[H^+]$ from 40 nM; or for every 10 mm Hg rise in PCO_2 from 40 mm Hg, expect a 3 mM rise in $[HCO_3]$ from 25 mM.

Respiratory Alkalosis

Acute: For every 1 mm Hg decrease in PCO_2 from 40 mm Hg, expect a 0.74 nM decrease in [H] from 40 nM; or for every halving of the PCO_2, expect a 2.5 mM fall in $[HCO_3]$.

Chronic: For every mm Hg fall in PCO_2 from 40 mm Hg, expect a 0.17 nM fall in [H]; or for every 2 mm Hg fall in PCO_2 from 40 mm Hg, expect a 1 mM fall in $[HCO_3]$ from 25 mM.

Essentials of Treatment

In mixed metabolic and respiratory acidosis, ventilate early, as this provides the greatest leverage to rapidly change the acid-base status.

When ventilating patients with chronic respiratory acidosis, ventilate them to their known chronic PCO_2.

Diagnostic Approach to Respiratory Alkalosis

In a patient with a low P_{CO_2}, begin by looking for a disorder associated with chronic respiratory alkalosis. If one is present, the patient is presumed to have chronic respiratory alkalosis and is assessed accordingly. On the other hand, if a chronic disorder is absent, the patient is assumed to have acute respiratory alkalosis.

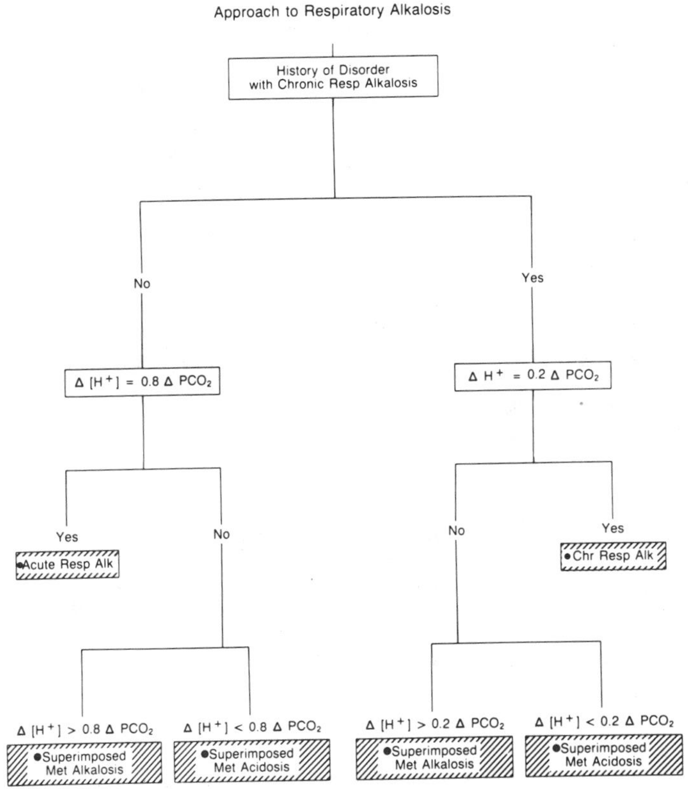

Flow Chart 14–H: Approach to the patient with respiratory alkalosis. The variables to evaluate are shown in the open boxes, and the final diagnoses are shown in the shaded boxes.

HYPONATREMIA

Definition

Serum [Na] < 136 mM.

Expected Normal Response

Type	Urine [Na]	Urine Volume	Urine Osmolality
Na loss	<10 mM	0.5 L/day	>800 mOsm
Water excess	>20 mM	= intake	<100 mOsm

Diagnostic Clues

Cause	Distinguishing Clinical Features
ADH release	
Volume depletion (Na loss type)	• Low serum osmolality • Urine [Na] < 10 mM
SIADH (Water excess type)	• Low serum osmolality, urea, and uric acid • Urine [Na] >20 mM
Hyperglycemia	• *Increased* serum osmolality and blood sugar
Hyperlipidemia	• *Normal* serum osmolality • Lipemia

Clinical Pearls

The danger is brain cell swelling.
Only CNS symptoms need aggressive therapy.
Symptoms are greater with acute fall in [Na].
Avoid rapid rise in [Na]; aim for + 1 mM per hour.

Essentials of Treatment (see Table 7–5, page 183)

Low Circulating Volume (Sodium Loss) Type

Stop water intake
NaCl if ECF volume-depleted.
Albumin if hypoalbuminemia.
Improve myocardial function if appropriate.
Dangers: If ADH secretion is turned off by therapy, water loss may be too rapid; limit with exogenous ADH.

Water Excess Type

Stop water intake.
Shift water from ICF with hypertonic saline. To raise the serum [Na] by 1 mM/hr, give 50 mmol Na per hour (in a 70 kg person, 42 ml of 5% saline) *plus* replace urine Na losses. May give furosemide to prevent ECF volume expansion.
Dangers: Congestive heart failure in elderly, and CNS damage with too-rapid increase in [Na].

QUESTIONS TO ASK WITH HYPONATREMIA

1. **Is the hyponatremia real?**
Answer: Hyperlipemia and hyperproteinemia did not cause hyponatremia if the plasma osmolality is low.

2. **Is hyponatremia due to hyperglycemia?**
Answer: In hyponatremia due to hyperglycemia, plasma osmolality is not low. For every 4 mM (72 mg/dl) rise in the plasma [glucose], expect close to a 1-mM fall in the plasma [Na].

3. **Is hyponatremia due solely to excess water intake?**
Answer: Almost never! If so, the urine will be maximally dilute and flow at 0.5–1 L/hr.

4. **Is hyponatremia due to high ADH levels?**
Answer: Almost always. Determine if there is an obvious reason for ADH being present. The usual causes are a reduction in "effective circulating volume," postoperative state, pain, emesis, or drugs that lead to ADH release; otherwise, the patient has the syndrome of inappropriate secretion of ADH (SISADH).

5. **What is the best way to treat hyponatremia?**
Answer: First, establish the cause of hyponatremia, because the treatments differ when the cause is a low "effective" circulating volume as compared with a water-excess type of hyponatremia.
Do not treat too vigorously; aim to raise the plasma [Na] 1 mM/hr. Only the symptomatic patient needs aggressive therapy.

HYPERNATREMIA

Definition

Serum [Na] > 145 mM due to either water loss or Na gain.

Expected Normal Response

Thirst

Urine: Minimal volume (0.5 L) and maximal concentration (osmolality > 900 mOsm/Kg H_2O).

Diagnostic Clues

Cause	Distinguishing Clinical Features
Renal water loss	• Urine volume > 5 L/day
Diabetes insipidus	• Urine Osm < 150 mOsm/kg
Central	• Responds to ADH with ↑ urine osmolality and ↓ volume
Nephrogenic	• Almost no response to ADH
Osmotic diuresis	• Urine Osm > 300 mOsm/kg H_2O
Sodium gain	• ECF volume expansion (very rare cause)

Clinical Pearls

Hypernatremia causes morbidity via brain cell shrinkage.
If thirst absent, basis is CNS lesion.
Avoid rapid lowering of [Na]; aim for −1 mM/hr.
IV solution must have lower [Na] than losses.
Cannot give D_5W faster than 300 ml/hr.

Essentials of Treatment

Stop ongoing water losses if possible (e.g., ADH).
Replace water deficit *and* ongoing losses.
Oral route for water is best.
Aim for correction at rate of −1 mM per hour.
250–300 ml water per hour will lower the [Na] by 1 mM per hour if no water losses.
Can assist correction by increasing urine Na loss with furosemide, and infuse Na as hypotonic saline.
Dangers: Too-rapid correction produces CNS damage.

QUESTIONS TO ASK WITH HYPERNATREMIA

1. **Is hypernatremia due to Na gain?**

Answer: Yes, if there is ECF volume expansion, thirst, and the excretion of the minimum volume of very concentrated urine. This is a rare clinical presentation.

A more common clinical example occurs during DKA when 150 mM saline is infused and the urine [Na] is < 100 mM.

2. **Has there been nonrenal water loss?**

Answer: If this is the case, the patient will have been thirsty and could not get to water; a minimum volume of maximally concentrated urine will be excreted.

3. **Is the cause a water shift into the ICF?**

Answer: This is extremely rare and requires a muscle problem in which the number of ICF particles has increased (mild rhabdomyolysis, convulsions). In this case, there will be no weight loss and an appropriate renal response, as described above.

4. **Is the cause excessive renal water loss?**

Answer: The 2 major groups of causes are diabetes insipidus (central or nephrogenic) or diuretic-induced water loss (usually osmotic owing to glucose or urea). The differential diagnosis is shown in the polyuria flow chart (13–A, page 289) and hinges upon whether the large volume of urine has an osmolality that is very low or in the 400–500 mOsm/kg H_2O range. In the case of diabetes insipidus, the central variety will have the appropriate history and a rise in urine osmolality 2–4 hr after ADH is given (not to maximal levels); in contrast, the nephrogenic variety will have no renal response to ADH and a renal disorder, or it will be secondary to lithium intake.

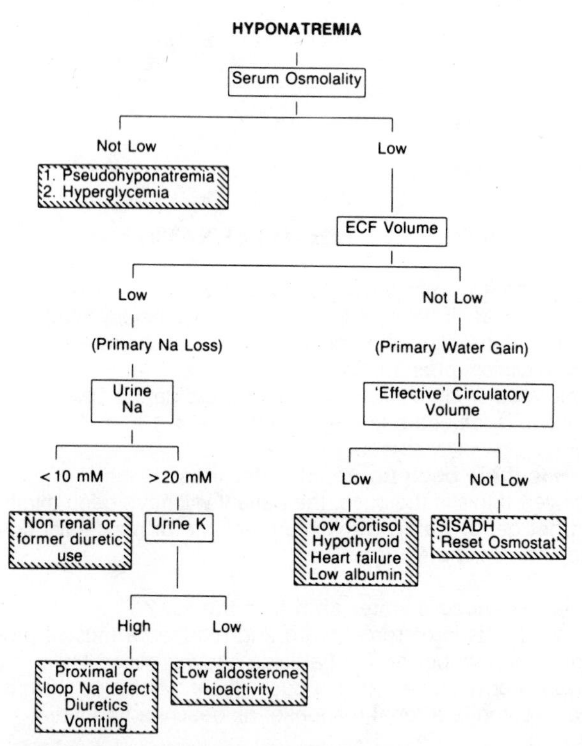

Flow Chart 14–I: Approach to hyponatremia. The clear boxes represent the variables to be evaluated clinically to reach a final diagnosis. The final diagnoses are shown in the shaded boxes.

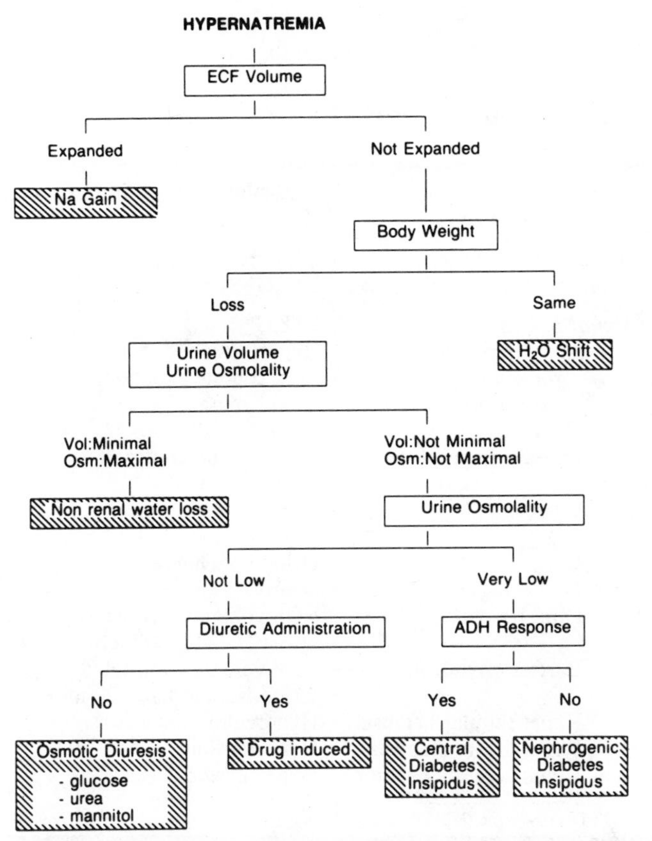

Flow Chart 14–J: Approach to hypernatremia. The open boxes represent the variables which, when evaluated, permit one to progress toward the diagnosis. The diagnostic groups are shown in the shaded boxes.

HYPOKALEMIA

Definition

Serum [K] < 3.5 mM.

Expected Normal Response

Urine K excretion < 30 mmol/day.

Diagnostic Clues

Causes	Distinguishing Clinical Features
Nonrenal K loss (e.g., diarrhea, laxative)	• Urine K < 30 mmol/day • TTKG* < 4
Renal K loss	• TTKG* > 7
Primary hyperaldosteronism or exogenous mineralocorticoid	• Hypertension • ECF volume expanded • Lower renin • Metabolic alkalosis
Secondary hyperaldosteronism	• ECF volume contracted • Elevated renin
Vomiting	• Urine Cl < 15; Na may be > 20 mM • Urine net charge may be positive
Diuretics	• ECF volume contracted • Urine Na and Cl fluctuate
Bartter's syndrome	• ECF volume contracted • Urine Na and Cl > 20 mM
Renal tubular acidosis	• Hyperchloremic acidosis
Renal artery stenosis	• Hypertension
Renin-secreting tumor	• Hypertension

*TTKG = urine [K]/U/P (Osm)/plasma [K]

Clinical Pearls

Only 2% of total body K is in ECF.
Abusers may not admit to their actions.
If [K] = 3 mM, deficit = 100–300 mmol.
If [K] < 2 mM, deficit > 600 mmol.
Avoid central K bolus whenever possible.
Must treat if diabetic ketoacidosis or on digitalis.

QUESTIONS TO ASK WITH HYPOKALEMIA

1. **Did a K shift into the ICF cause hypokalemia?**
Answer: This possibility is unlikely to cause severe hypokalemia.

2. **What features should be stressed on history?**
Answer: Diuretics, vomiting, and laxative abuse are common causes, but the patient may deny them.
If digitalis is being used, the danger of arrhythmias is much greater.

3. **What should be stressed on physical exam?**
Answer: ECF volume, hypertension, and psyche.

4. **What should be stressed on lab exam?**
Answer: ECG, blood acid-base status, urine [K], [Cl], [Na], flow rate, osmolality, and the TTKG.

5. **When is hypokalemia most dangerous?**
Answer:
(a) In combination with digitalis;
(b) When the K deficit is severe or sudden;
(c) When associated with metabolic acidosis, especially DKA, because therapy will unmask the severe K deficit;
(d) Overaggressive IV K therapy producing transient or severe hyperkalemia.

6. **How much K should be given during therapy?**
Answer: Usually several hundred mmol.
(a) Use the oral route if possible.
(b) IV K must not be given too quickly or in large central boluses.
(c) If IV, use several routes, and do not exceed 60 mmol/hr.
(d) If correcting ECF volume contraction with saline, consider amiloride to avoid excessive renal K loss.
(e) *Dangers:* Too-rapid K administration may result in arrhythmias. Avoid central K boluses unless emergency.

HYPERKALEMIA

Definition

[K] > 5.5 mM.

Expected Physiologic Response

Renal K excretion > 100 mmol/day.

Diagnostic Clues

Cause	Distinguishing Features
K load	• High K excretion; TTKG > 7
K shift from ICF	• High K excretion; TTKG > 7
Low renal K excretion	
↓ Na delivery	• Urine [Na] < 10 mM
↓ Mineralocorticoid bioactivity	• Urine [Na] > 20 mM • TTKG < 5

Clinical Pearls

Thrombocytosis can cause factitious hyperkalemia.

Hyperglycemia in a diabetic may cause severe hyperkalemia if hypoaldosteronism is also present.

Major threat of hyperkalemia is arrhythmias.

Only 2% of total body K is in ECF.

EKG changes indicate urgent need for therapy.

Treat [K] > 6.0, even without EKG changes.

Essentials of Treatment

STOP K intake.

If EKG changes, antagonize with 10 ml Ca gluconate.

Promote K shift into cells with 500 ml 10% glucose containing 10 units of regular insulin and 100 mmol of $NaHCO_3$.

Remove K via GI tract, using K-binding resins by enema (100 gm) for more rapid effect and by mouth (30 gm with 20 ml 70% sorbitol) for more long-term control.

Reverse renal defect if possible (e.g., increase urine volume or Na delivery).

If ongoing K load or if severe hyperkalemia, arrange for dialysis (hemodialysis more efficient than peritoneal).

QUESTIONS TO ASK WITH HYPERKALEMIA

1. **Is hyperkalemia due to a lab problem?**
Answer: Rule out hemolysis or thrombocytosis. ECG changes exclude a lab problem.

2. **Is hyperkalemia due to high K intake alone?**
Answer: Not likely; high K intake will contribute only if K excretion is compromised.

3. **Did K shift from the ICF?**
Answer: Yes, if there is cell damage, insulin, β-adrenergic or aldosterone blockade or deficiencies, or acute metabolic acidosis. Rarely, drugs will cause depolarization and a K shift from the ICF.

4. **Is K excretion compromised?**
Answer: K excretion is determined by the urine [K] × urine flow rate; therefore, decreased K excretion will occur if there is a problem with aldosterone or Na delivery (low urine [K]) or low urine flow (renal failure, low ECF volume).

5. **When is hyperkalemia an emergency?**
Answer: If there are ECG changes, an anticipated rise in the plasma [K], or a plasma [K] $>$ 7.0 mM.

6. **What should be stressed on history?**
Answer: Diet K, catabolism, diabetes mellitus, drugs interfering with K excretion (see Table 11–2, page 263). The symptoms of adrenal insufficiency may be nonspecific.

7. **What should be stressed on physical exam?**
Answer: ECF volume status, skin pigmentation.

8. **What should be stressed on lab exam?**
Answer: ECG, blood acid-base status, glucose, urea, urine electrolytes, TTKG.

9. **What are the main concerns during therapy?**
Answer: Eliminate K intake. Shift K into cells with insulin and $NaHCO_3$. Improve renal K excretion if urine [Na] low and can be raised or aldosterone action instituted. Promote GI K loss with resins. If severe hyperkalemia, ECG changes or an acute [K] rise, prepare to dialyze and give Ca to antagonize cardiac effects of hyperkalemia.

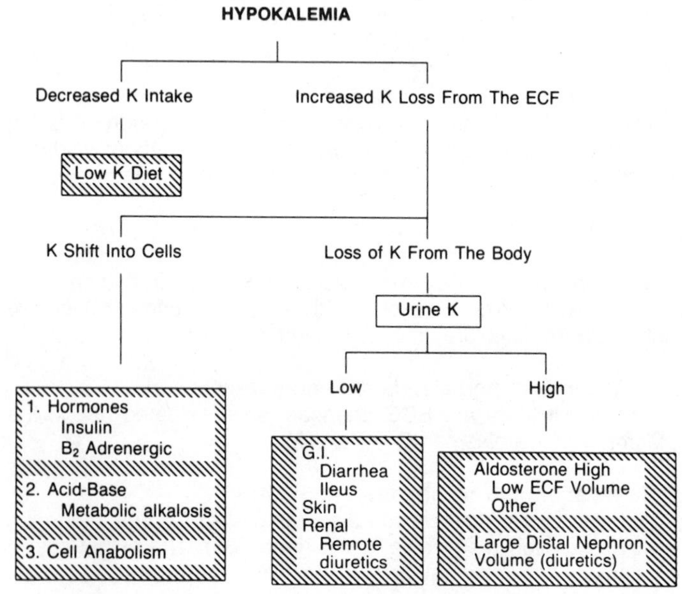

Flow Chart 14–K: Approach to hypokalemia. The open box represents the variable which, when evaluated, permits one to proceed to the diagnosis. The final diagnostic groups are shown in the shaded boxes.

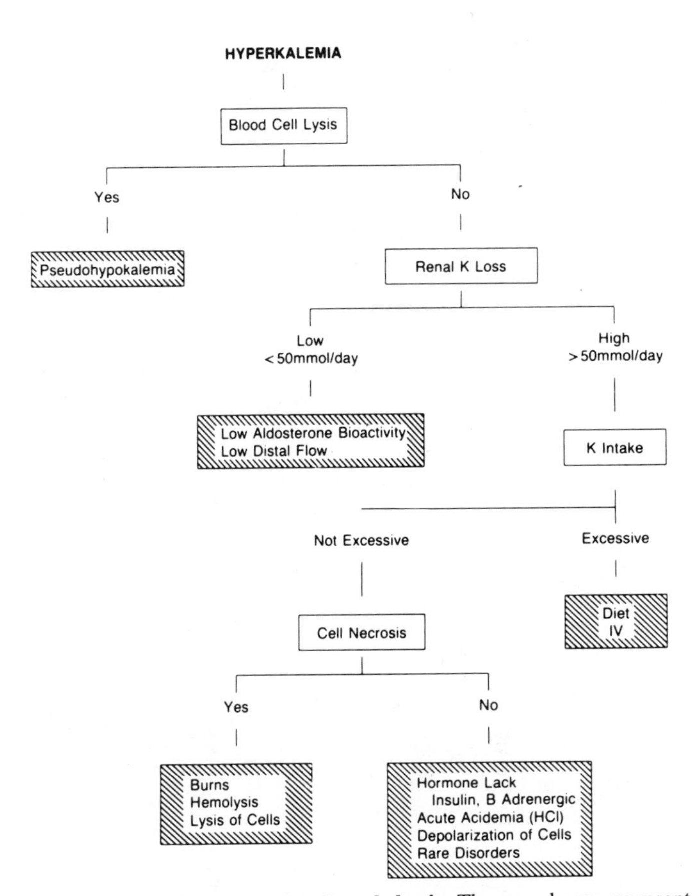

Flow Chart 14–L: Approach to hyperkalemia. The open boxes represent the variables that, when evaluated, permit one to proceed to the diagnosis. The final diagnostic group is shown in the shaded boxes.

INDEX

Note: Page numbers in *italics* refer to figures and flow charts; numbers followed by (t) refer to tables.